A HEAD START ON LIFE SCIENCE

Encouraging a Sense of Wonder

A HEAD START ON LIFE SCIENCE

Encouraging a Sense of Wonder

William Straits

Arlington, Virginia

Claire Reinburg, Director
Rachel Ledbetter, Managing Editor
Deborah Siegel, Associate Editor
Andrea Silen, Associate Editor
Donna Yudkin, Book Acquisitions Manager

ART AND DESIGN
Will Thomas Jr., Director
Joe Butera, Senior Graphic Designer, cover and interior design

PRINTING AND PRODUCTION
Catherine Lorrain, Director

NATIONAL SCIENCE TEACHERS ASSOCIATION
David L. Evans, Executive Director
David Beacom, Publisher

1840 Wilson Blvd., Arlington, VA 22201
www.nsta.org/store
For customer service inquiries, please call 800-277-5300.

21 20 19 18 4 3 2 1

NSTA is committed to publishing material that promotes the best in inquiry-based science education. However, conditions of actual use may vary, and the safety procedures and practices described in this book are intended to serve only as a guide. Additional precautionary measures may be required. NSTA and the authors do not warrant or represent that the procedures and practices in this book meet any safety code or standard of federal, state, or local regulations. NSTA and the authors disclaim any liability for personal injury or damage to property arising out of or relating to the use of this book, including any of the recommendations, instructions, or materials contained therein.

Library of Congress Cataloging-in-Publication Data

Names: Straits, William, author.
Title: A head start on life science : encouraging a sense of wonder / by William Straits.
Description: Arlington, VA : National Science Teachers Association, [2017] | Includes bibliographical references.
Identifiers: LCCN 2017040954 (print) | LCCN 2017043635 (ebook) | ISBN 9781681403496 | ISBN 9781681403489
Subjects: LCSH: Life sciences--Education (Elementary) | Nature--Education (Elementary)
Classification: LCC QH315 (ebook) | LCC QH315 .S865 2017 (print) | DDC 507.1--dc23
LC record available at *https://lccn.loc.gov/2017040954*

CONTENTS

Plants

4

Nature Walks

Appendixes

About the Author

William Straits is a professor of science education at California State University, Long Beach. He earned a bachelor's degree in biological sciences from the University of California at Irvine (1991); master's degrees in biology from California State University Fullerton (1995) and in curriculum and instruction from the University of Texas at Austin (1997); and a PhD in science education also from the University of Texas at Austin (2001). Throughout his career, as a science teacher and as a teacher educator, he has emphasized the natural and important connections between science and language literacies. He currently serves as director of the National Center for Science in Early Childhood and focuses much of his scholarly work on early childhood science education. He is continually amazed by the loving and tireless work of teachers of young children and humbly hopes that this book, in some small way, aids them in their efforts—helping children to heighten and expand their joyful "sense of wonder" about the natural world.

Contributors and Collaborators

The following science and early childhood educators made significant contributions to the lessons presented here. They are considered as coauthors of the lessons they helped develop, and their names appear at the beginning of the lessons.

Susan Gomez Zwiep
Professor, Science Education
California State University, Long Beach
Long Beach, CA

Angelica Gunderson
Science Teacher
Norwalk-La Mirada School District
Norwalk, CA

Nicole Hawke
First-Grade Teacher
Coachella Valley Unified School District
Thermal, CA

Emily Kraemer
Child Development Specialist
Orange Coast College, Harry and Grace Steele Children's Center
Costa Mesa, CA

Myra Pasquier
Science Teacher
Montebello Unified School District
Montebello, CA

Lauren M. Shea
Education Lecturer
American University
Washington, DC

Kristin Straits
Science Education Lecturer
California State University, Long Beach
Long Beach, CA

We would also like to recognize the educators who helped contribute to this text by reviewing or pilot testing the lessons.

Hope Gerde
Associate Professor, Human Development and Family Studies
Michigan State University
East Lansing, MI

Nicole Hawke
First-Grade Teacher
Coachella Valley Unified School District
Thermal, CA

Mary Lopez
Preschool Teacher
California Heights Children's Center
Long Beach, CA

Ruth Piker
Professor, Early Childhood Education
California State University, Long Beach
Long Beach, CA

Jennifer Robinson
Early Learning Specialist
Buena Park School District
Buena Park, CA

Kim Watten
First-Grade Teacher
Long Beach Unified School District
Long Beach, CA

R. Russell Wilke
Professor and Chair, Biological Sciences
Angelo State University
San Angelo, TX

Additionally, we would like to recognize the following contributors who helped develop the overall format of the lessons.

Hope Gerde
Associate Professor, Human Development and Family Studies
Michigan State University
East Lansing, MI

Bradley Morris
Associate Professor, Educational Psychology
Kent State University
Kent, OH

Christina Schwarz
Associate Professor, Teacher Education
Michigan State University
East Lansing, MI

Laurie Van Egeren
Director, University Outreach and Engagement
Michigan State University
East Lansing, MI

Finally, we would like to thank the children, families, and amazing staff at the Harry and Grace Steele Children's Center at Orange Coast College for allowing us to photograph the lessons in action.

Science for Young Children

Our Theme

In 1956, marine biologist and conservationist Rachel Carson wrote a book, *The Sense of Wonder,* about the time she spent along the Maine coastline with her young nephew, Roger. From the time Roger was just a baby until he was more than 4 years old, he and Rachel shared adventures in the world of nature. She never set out to "teach" him anything, but rather to have fun and marvel at the plants and animals, the sounds and smells, the rocks and waves they encountered on walks through the woods and along the ocean. Roger, of course, learned a great deal as Rachel explored with him, calling his attention to various things and talking with him about what they observed. Roger learned as the two of them made discoveries *together*.

In *The Sense of Wonder,* Carson wrote, "a child's world is fresh and new and beautiful, full of wonder and excitement. It is our misfortune that for most of us that clear-eyed vision, that true instinct for what is beautiful and awe-inspiring, is dimmed and even lost before we reach adulthood. [I wish that] each child in the world be [given] a sense of wonder so indestructible that it would last throughout life …" (Carson 1956). "Sense of wonder" has become the theme of our *A Head Start on Science* (*HSOS*) program (see Appendix C); we strive to bring teachers of young children the resources and support they need to heighten and expand children's innate curiosity about the natural world. *A Head Start on Life Science* lessons are written to help adults facilitate young children's learning as they work as partners in exploring the natural world. We hope your sense of wonder will be heightened as you engage in science explorations with children, actively following as their curiosity leads them to discoveries about all that they see, hear, smell, and touch.

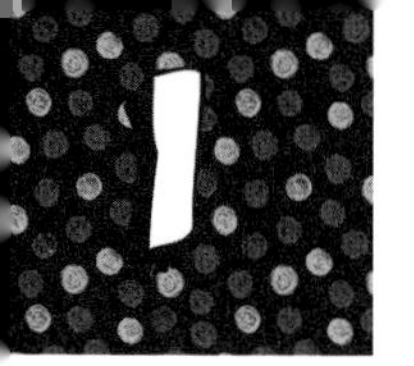

High-Quality Early Childhood Science Education

Science learning experiences are important in early childhood education. Early childhood science education (ECSE) engages teachers and children in high-quality interactions that can not only promote children's understanding of science concepts and skills but also can narrow the achievement gap (Cabell et al. 2013) and provide a meaningful context for developing literacy and math skills (Gelman et al. 2009). In fact, compared to other learning contexts (e.g., reading or math instruction), teachers engage in higher-quality teacher–child interactions when they engage in science (Cabell et al. 2013). These high-quality interactions, including supports for concept development, expanding on child ideas, and use of open-ended questions and advanced language, significantly enhance children's cognitive development and thus their academic outcomes (Mashburn et al. 2008). Additionally, ECSE can lead to gains in language achievement for English-language learners, particularly for speaking and listening skills (Gomez Zwiep and Straits 2013).

However, ECSE varies greatly in classroom practice. In the extreme cases, "science" experiences look more like arts and crafts projects (for example, painting pumpkins or gluing feathers on paper birds) that do little to promote children's science learning. At the opposite extreme, ECSE can consist of teacher-directed activities that emphasize academic learning and scientific vocabulary. Although the balance in ECSE has been much debated, this book is written with the belief that the optimal early learning experiences for children lie somewhere between these two extremes. ECSE should not be a series of isolated activities that occupy children but fail to engage them in prolonged investigation or produce long-lasting, meaningful learning. Likewise, ECSE should not consist of experiences that are entirely teacher directed and academic, placing emphasis on the products of learning (e.g., vocabulary) rather than the process; that do not develop children's abilities to engage in science practices; and that fail to foster science dispositions such as persistence, curiosity, questioning, and exploring.

ECSE experiences should address topics relevant to children's everyday experiences that can be experienced firsthand, serve as bases for collaboration and communication among children and adults, and have the potential for meaningful investigation. And, although these high-quality ECSE experiences can take many different forms, they generally have three components at their core: generating and relating to children's interests; facilitating collaborative, child-driven investigation; and providing opportunities for children to reflect on, represent, and apply what they've discovered. The three components are sequential, systematically building children's understanding. Importantly, the three components should be accomplished over days of instruction, in amounts of time consistent with the development and age of your children.

High-quality ECSE experiences begin with children's interest and curiosity. At the beginning of ECSE learning experiences, teachers must tap into children's existing curiosity and generate new interest about the phenomenon to be explored. Aligning with children's curiosity helps to ensure that children are intellectually engaged and sustain interest during a sequence of prolonged and meaningful investigations of science phenomena present in the world around them. Although teacher facilitated, these investigations need to be child driven, emphasizing children's decisions and meaning-making. It is during investigations that children develop their skills in using science practices, such as observing, measuring, comparing, sorting, communicating, and graphing, as well as important language and social skills. Although engaging in science investigations and employing different science and communication skills are important for learning, they alone are insufficient. For meaningful, lasting learning, children also need opportunities to reflect on, represent, and apply their new understandings. This thinking about, sharing, and using new understandings helps children to solidify what they've learned, builds metacognitive skills, and also leads children to new explorations. The lessons in *A Head Start on Life Science* are consistent with this view of learning and are written to help you as the teacher learn to design and implement effective science learning experiences.

Our Beliefs About Science for Young Children

Central to our understanding of young children is the idea that a sense of wonder is innate; children are naturally awestruck by and curious about the natural world. Further, the exploration of this amazing natural world is natural for young learners. We believe that a sense of wonder is part of all children's experience and that children are intrinsically motivated to explore the natural world. Therefore, it is important that *all* children have access to culturally relevant science experiences that are of value in learners' everyday worlds. Formal science education settings must tap into this natural interest in science by providing authentic materials, allowing a degree of child autonomy, and celebrating each child's success. Additionally, science education for young learners must utilize play and emphasize free exploration as a means for learning, provide opportunities to teach and learn from peers, recognize that trial and error are natural parts of the scientific learning experience, and emphasize the importance of process over right and wrong answers.

Teachers charged with facilitating science education for young learners face a great challenge. Teachers must abandon the traditional view of the teacher as disseminator of information and adopt roles as facilitators of learning. Consistent with this, a primary role of the early childhood science teacher is to provide an appropriate learning environment and opportunities for children to explore, represent, and share their discoveries. Teachers need to model excitement and enthusiasm when involved in science exploration and when planning and anticipating discoveries. Throughout the design of learning experiences, teachers need to recognize that the process of discovery and the science practices children engage in are more important than learning science facts and that science experiences can be highlighted at all times and in all parts of the classroom and outdoors, not just during "science time" or at the "science center." As children engage in science investigations, adults should observe children's actions and listen to children's conversations so that they can follow children's leads; child-initiated learning is of great importance and should be encouraged and supported. Additionally, effective early childhood teachers must be effective parent educators and involve families in their children's science activities.

Throughout our work with children we must all emphasize the exciting process of discovery over science information. Rather than understanding science as the learning of already-known answers, our children should see science as the exploration of a vast and wondrous world of infinite mysteries. The experience is the objective; instead of telling children science facts, nurture their curiosity, interest, and joy. These attitudes and foundational experiences can serve as a basis for a lifelong love of science.

We firmly believe that children enrolled in the types of early childhood programs described here,

where active learning and children's exploration are central, where teachers emphasize science practices and child-initiated investigation, and where teachers and families are actively involved in children's science learning, are more likely to succeed in school, and in life, than children who are denied these important learning experiences.

Developmentally Appropriate Science

The lessons in *A Head Start on Life Science* were created with these beliefs in mind and designed for developmentally appropriate use in early childhood education settings. Inspired by and adapted from the activities in NSTA's *A Head Start on Science: Encouraging a Sense of Wonder,* edited by the founder of our project, William Ritz, each lesson has as its basis active, hands-on involvement of children and focuses not on teaching children "science facts," but rather on nurturing children's innate curiosity about the natural world and encouraging children to make discoveries on their own. Teachers familiar with the activities in the original book will find the lessons here to be useful models for expanding science activities into integrated inquiry lessons. The lessons are intended to help teachers to expand children's thinking in an area of interest and are designed to help teachers to come to understand a method for sequencing learning opportunities that promotes understanding. In ECSE, children's interest and prior knowledge are key—developmentally appropriate science meets children where they are and allows children to participate at their own level. These lessons provide experiences that are hands-on, concrete, and relevant and allow children to learn through play and social interaction. *A Head Start on Life Science* integrates learning across domains, allowing for science learning throughout a child's school day as well as at home with his or her family. The learning situations across all lessons and activities are flexible, allowing teachers to follow children's interests and questions. And throughout all the lessons we emphasize children's thinking and use of science practices rather than focus on factual knowledge or right or wrong answers.

Lesson Overview

Introduction

Each lesson in *A Head Start on Life Science* provides teachers with introductory information to help prepare for the learning experience. The information includes a brief description of the lesson and a listing of the learning objectives, required materials, and safety considerations. This information is here to assist you in preparing your lesson and deciding where in your unit of study this lesson would best fit. Additionally, for each lesson we provide relevant teacher content background. In our many years of working with early childhood educators, we have often heard teachers express a desire for more science content. We are providing it in this revised version. However, we emphasize that this information is for teachers' knowledge only and is there to assist you in your own learning. This information should NOT be directly taught to young children. Studies have shown that formal, teacher-directed instruction is at best ineffective and at worst detrimental to children's long-term growth as learners. Our goal is not for children to acquire "facts," but to be active explorers, reveling in the process of discovering more about the natural world around them. In addition to the teacher content background, we provide a list of key science terms for teachers to be aware of as they prepare for the lesson. Keep in mind that vocabulary acquisition is not our primary intent. Capitalize on opportunities for vocabulary development when they arise naturally during your conversations with children, but do not feel compelled to force vocabulary into these learning situations. In all situations, our primary goal is to enhance children's sense of wonder—their innate curiosity about and appreciation for the natural world all around them. Following the introductory material, the procedure for each science lesson is described.

Procedure

Aligned with high-quality ECSE and inspired by the learning cycle[1] (Atkin and Karplus 1962), we've designed lessons where children are first, oriented toward the topic to be investigated; second, given a chance to explore and develop an understanding of concept; and third, supported in formalizing that understanding by explaining or applying. Subsequently, the procedure for each lesson is divided into three sections: *Getting Started*, *Investigating*, and *Making Sense*, each of these sections serving an important purpose in the development of new knowledge. Although there is broad consistency across lessons, the specific ways that children are oriented toward a topic, explore and develop an understanding, and formalize that understanding can vary from lesson to lesson. We have been intentional in our effort to highlight these alternatives across the lessons provided here, and these options are described below and in the model in Figure 1.1, on page 6.

In all learning, it is important to give children an introduction before engaging in teacher-structured explorations. We call this introduction *Getting Started* and encourage teachers to use this time to allow children to play with materials, share what they already know about a topic, and ask questions. Only after this introduction are children ready for directed exploration. *Getting Started* helps prepare children to learn by activating prior knowledge, allowing free exploration of materials, generating interest, and giving purpose to their investigations. What you choose to do with children can vary from lesson to lesson, but for all lessons children need an opportunity to orient to the topic before structured learning experiences can occur. There are five categories of teacher actions in the *Getting Started* section; most lessons employ two or three of these. The teacher actions are

1. **Prior knowledge.** Show or describe an example of the topic of interest and use questioning to probe children's prior knowledge about the topic. (For example, "What do you know about elephants?")
2. **Introduction.** Introduce phenomena or materials to children to explore, asking children to describe them. (For example, "Look closely at these seeds and share what you notice.")
3. **Child questions and curiosity.** Ask what they know about the phenomenon and what questions they have about it. (For example, "What questions do you have about fish?")
4. **Prompting questions.** After children have gotten a chance to explore and wonder themselves, introduce a question. (For example, "How have the seeds changed?")
5. **Initial explanations.** Ask children to explain the reasoning behind their initial answers to your prompting questions and record their responses. (For example, "Why do you think it will move or change that way?")

1 The learning cycle is a three-part teaching model. The first phase, "exploration," has students engaged in teacher-facilitated experiences; in the second phase, "concept development," teachers guide students in understanding the concept(s) related to their explorations; and finally, in "concept application," the teacher presents a situation for students to use their new understanding(s). In this book, you'll find parallels between "exploration" and the *Investigating* of many lessons, as well as between the "concept development" and "concept application" phases and some aspects of *Making Sense*.

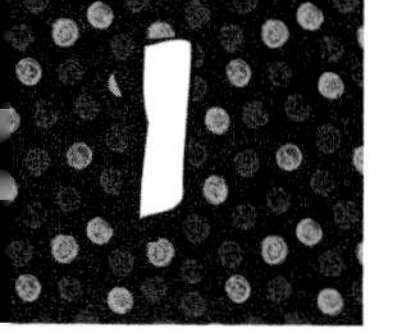

Figure 1.1

Lesson Model

Model demonstrating the three-phase lesson sequence used for each of the A Head Start on Life Science lessons and the components that can be included in each of the three phases.

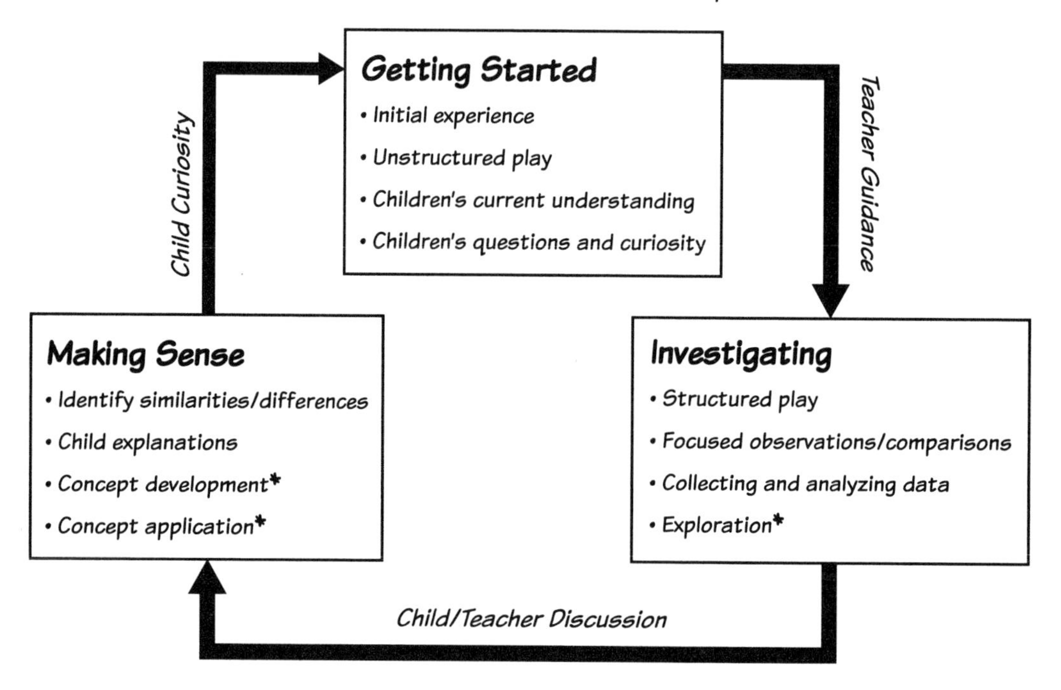

* Denotes connections to the learning cycle (Atkin and Karplus 1962)

All five do not need to be in each lesson, but high-quality science lessons nearly always begin with one (or more) of these actions.

Investigating provides an opportunity for children to actively work with materials to generate new understandings of and appreciation for some natural phenomenon. Capitalize on opportunities to discuss new science concepts and vocabulary, but use a light touch. With our young children, the experience is our primary objective. Let children's interests shape the direction and outcomes of your science explorations. The *Investigating* descriptions (like all of the descriptions) provided in each lesson are intended as guides, suggesting one set of possibilities. You do not need to adhere strictly to them. Let your children's curiosity take you and your lesson in new directions. Throughout this time, use questioning to encourage children to observe details, make comparisons, and notice differences that might otherwise have been overlooked. Keep in mind that your aim in this questioning (and in questioning throughout the lesson) is to be conversational; be a learner with your children. Share ideas, insights, excitement, and questions. Model for your children what enthusiastic science learning looks like. Whether you are a preschooler or a professional scientist, there are two important parts to investigating:

2. **Collecting data.** This includes experimenting, observing, measuring, and documenting and can be encouraged by the teacher by asking prompting questions such as "How can we test that idea? What do you notice? How big is it? How could we keep track of these ideas?"
3. **Analyzing data.** This includes comparing and sorting (across objects and processes as well as across children's ideas and findings), encouraged by prompting questions such as "Which of these are similar? Which are different? How are these alike? In what ways are these different? Can you put these in order from smallest to largest?"

In the *Investigating,* phase children take center stage as they actively explore phenomena. However, teachers play a very important role in these investigations, including providing appropriate materials and supports; sequencing activities so that they build understanding and curiosity; and being active participants, engaging in explorations with children. The conversations and questions you have with children will help them to better understand and appreciate the science topic being explored. [Note: Sometimes lessons have two separate but related activities that take place during the *Investigating* section. When this occurs, the individual activities are designed to be experienced sequentially and are referred to as *Investigating 1* and 2.]

Making Sense provides children with an opportunity to represent and share results and to apply, sum up, think about, and discuss what they discovered during their investigations. To develop critical thinking skills and a more lasting understanding of what they learn, learners need to be given the opportunity to reflect on their investigations. This reflection may mean the discussion of similarities or differences or the identification and generalization of patterns discovered during the *Investigating* section. Teacher questioning and guidance are important to facilitate this sense making. Throughout the *Making Sense,* section, remember that child explanations, based on discoveries made during their Investigating, are more important than the "correct" scientific explanation. There are four important processes in *Making Sense:*

1. **Describing findings.** Articulating our thoughts helps us to clarify and reinforce our understandings; this is true for children as well as adults. Communicating about the outcome of an exploration is important for learning and can be part of the investigation itself or the sense making that follows. In *Making Sense,* encourage children to describe and represent (through drawings, text, graphs, etc.) what they've learned and how different phenomena compare. For example, while making sense of an investigation, children might state, "The spiny seed sticks to things, but the smooth seed doesn't" or "All of these birds have feathers."
2. **Generating explanations and identifying patterns.** Here, learners, just like scientists, take an investigation one step further, going beyond simply stating the results to explaining the meaning of the results or the trend across multiple results. In these efforts, children use (and further develop) their reasoning and critical-thinking skills as they engage in concept development. For example, after exploring the insides of several fruits, a child may be able to tell you that (or ask you if) all fruits have seeds. Although developmentally, many of our younger children may not be ready to make these leaps, as a part of science investigations they should still have opportunities to explain their thinking at a level appropriate

for each individual child. The correct explanation is not our goal; our goal is to provide children with opportunities to explain their thinking and develop their reasoning skills.

3. **Application.** Opportunities to apply new knowledge are important for long-term learning. In *Application* children are asked to use what they discovered during the investigation(s) to explore a different, yet similar, phenomenon. For example, after developing the skill of sorting using beans, children could use this new skill to sort buttons; after studying a goldfish, children could observe other fish to see if they too have fins, gills, tails, and so on.

4. **Curiosity**. Science investigations often lead to new ideas and new investigations. This continued exploration is one of the aspects of science that most interests scientists and motivates them to devote their lives to scientific study. We want to help nuture this same passionate curiosity in our children by encouraging them to identify what questions they have about the phenomenon studied and what they'd like to explore next. Let their interests determine your next *Getting Started* experience.

Each lesson encourages children to engage in one or more of these four processes. And very often, these processes will prompt the exploration of new topics or questions; a successful lesson is one that inspires further learning. [Note: As in the *Investigating* section, on occasion there are two parts to this final section. When this happens, the parts are referred to as *Making Sense 1* and 2.]

Beyond the Lesson

After the lesson, we provide ideas for extending the science learning in a section called *What's Next?* We first provide an extension activity that describes one or more science activities related to the topic explored during the lesson. It's our hope that these extension activities link to the *Making Sense* section by connecting to children's curiosity and offering additional opportunities for children to apply what they've learned. The ideas provided in the extension activity are not as detailed as in the lesson; it's up to the teacher to structure these activities in a manner consistent with the *Getting Started*, *Investigating*, and *Making Sense* format.

It's important to remember that science need not be a stand-alone subject limited to science time or relegated to a dusty science table. Science naturally connects to math and language arts. Ideas for making these connections are provided in the *Integration to Other Content Areas* section. In *Other Connections*, we present still more ideas for connecting the science lesson to other parts of the school day. Examples for increasing the relevance of the science lesson in the child's life are given, as are ideas for the art, sensory, and dramatic play centers. Additionally, the Family Activities section provides ideas for how families might interact with their child; these are provided in English and Spanish and are written with parents and guardians in mind.

What to Look For and *Standards* are the final sections provided. The *What to Look For* section lists questions that teachers can ask themselves in an attempt to gauge children's progress toward learning goals. After each question, the academic skills that may be demonstrated by children as they respond are briefly indicated. These academic skills are based on indicators included in several preschool and primary grade frameworks and assessments,including HighScope's *Preschool Child Observation Record* (2015), *Teaching Strategies GOLD: Objectives for Development and Learning* (Burts et al. 2016), *Head Start Early Learning*

Outcomes Framework (U.S. Department of Health and Human Services, 2015), NAEYC Early Childhood Program Standards and Accreditation Criteria (2015), and the *Next Generation Science Standards (NGSS)* Science and Engineering Practices (2013). Use children's responses to these questions to make decisions about your next steps in instruction, providing additional experiences for children who have yet to demonstrate mastery of each goal. Be sure to consider each child's age and development to ensure that the tasks related to each goal are appropriate. Finally, for each lesson we provide the most relevant standards from the Head Start Early Learning Outcomes Framework, the science and engineering practices (SEP) of the *NGSS* (NGSS Lead States 2013), and the kindergarten or first-grade *Common Core State Standards for Mathematics and English Language Arts* (NGAC and CCSSO 2010). Although each lesson has multiple points of alignment with these different standards documents, only the primary standard addressed in each is noted and detailed at the end of each lesson. [Note: These different standards documents point out an important consideration for teachers using the lessons presented here. *A Head Start on Life Science* is written for teachers of children ages 3–7. This is an enormous span considering the development of children. You will have to be a critical reader of these lessons, making sure that the suggestions are developmentally appropriate and applicable to your setting. Each component of each lesson, while striving to adhere to the same general format, is written to present varied tasks to children. Pick and choose from among them and adapt them as necessary to design the optimal learning experiences for your children. For example, Head Start teachers may want to design lessons that emphasize the Cognitive (Science and Math) or Literacy Domains as suggested in the Head Start Early Learning Outcomes Framework (see Appendix A for a guide). Elementary teachers may want to make their science lessons two- or three-dimensional, aligning with a combination of the SEPs, disciplinary core ideas, and cross-cutting concepts of the *NGSS* (see Appendix B for a guide). Providing for each of our many different audiences *within* all aspects of each lesson is not practical, so instead we offer ideas for pre-K and Elementary (and for rural, urban, and suburban settings, Spanish- and English-speaking families, etc.) *across* the lessons. Our hope is that as a whole the lessons here are "educative," and that as you read and try the lessons you'll learn more about what makes for appropriate, high-quality science for children.]

Planning for a Lesson

Our lessons include several components. These are provided so that you may provide multiple and different opportunities for students to explore each science phenomenon. However, it is not our intention that these lessons be necessarily taught in full or all at once. Depending on the development of your children, a different number of activities, each of a different duration, may be most appropriate. As you plan learning activities for your children, consider the sample weeklong plans provided in Tables 1.1 and 1.2 (p. 10).

Planning for Safety

With hands-on, process- and inquiry-based classroom and field activities, the teaching and learning of science can be both effective and exciting. However, successful science teaching must always address potential safety issues. Teachers should review and follow local polices and protocols used at their school site and within their school district or agency (e.g., Board of Education safety policies, field trip policies, etc.). Additional applicable standard operating procedures can be found in the National Science Teacher Association's

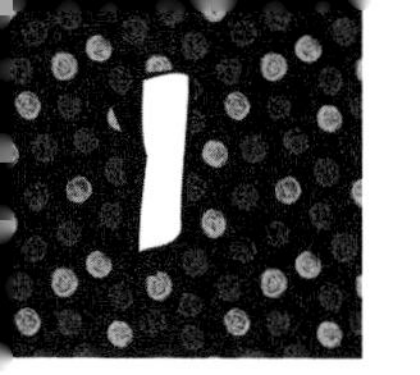

Table 1.1

Weekly Plans for Children Ages 3 to 5

Monday	Tuesday	Wednesday	Thursday	Friday
• Getting Started • Book From Reading Connection • Centers	• Investigation • Book From Reading Connection • Centers	• Making Sense • Book From Reading Connection • Centers	• Extension Activity • Book From Reading Connection • Centers	• Math and/ or Writing Connections • Centers • Send Home Family Activity

Monday	Tuesday	Wednesday	Thursday	Friday
• Getting Started • Centers	• Investigation • Centers	• Making Sense • Book From Reading Connection	• Extension Activity • Book From Reading Connection	• Math and/ or Writing Connections • Send Home Family Activity

Table 1.2

Weekly Plans for Children Ages 5 to 7

Monday	Tuesday	Wednesday	Thursday	Friday
• Science Lesson • Book From Reading Connection • Send Home Family Activity	• Lesson Review • Vocabulary Review (if appropriate) • Book From Reading Connection	• Extension Activity • Book From Reading Connection	• Math and Writing Connections	• Centers

"Safety in the Science Classroom, Laboratory, or Field Sites" document, available online at *www.nsta.org/docs/SafetyInTheScienceClassroomLabAndField.pdf*. Remember that children with allergies or immune-system illnesses will need additional consideration. Always check with parents and guardians and the school nurse ahead of time. Field studies need to be given special consideration. As with all field trips, scout out the area before you take children there. During all science activities, but especially during field studies, make sure there is appropriate student supervision by adults. For information on field trip safety, read the NSTA Safety Advisory Board paper "Field Trip Safety." It can be found at *static.nsta.org/pdfs/FieldTripSafety.pdf*. [Note: The safety precautions provided in each lesson are based in part on use of the recommended materials and instructions, legal safety standards and better professional practices. Selection of alternative materials or procedures for these activities may jeopardize the level of safety and therefore is at the user's own risk. Ultimately, teachers are responsible for the materials and procedures they opt to use in their classrooms and for the safety and well-being of their children.]

2

Animals

The components of the natural world that children often find most fascinating are animals. This interest, coupled with their great diversity and their prominence in our communities and cultures, makes animals the ideal topic for exploration with young children. As you investigate animals together, keep in mind that children have been fascinated by and thinking about animals long before they came to your classroom. In all of your lessons, make time for children to share what they know about animals and what they wonder about animals. Listen to your children and follow their leads. The lessons in this chapter will give you and your children opportunities to explore several animals and to learn about animal structures and their functions, adaptations such as camouflage, and the diversity of life.

Throughout your explorations with animals, always keep safety in mind—both the safety of your children and the safety of the animals! Familiarize yourself with NSTA's guidelines for the responsible use of animals in the classroom, found online at *www.nsta.org/about/positions/animals.aspx*. After each lesson, return animals to their rightful place. If you collected animals from your yard, return them to your yard. If you purchased them from a pet store, return them to the pet store. Animals obtained from stores should never be released outside. In many cases, these non-native species can cause great harm to native species and existing natural systems as well as to crops and gardens. During lessons, engage children in conversations about the appropriate handling and treatment of animals and encourage your children to be kind to animals. Remember that throughout these lessons, your gentle and respectful treatment of animals will serve as an important model for our young children.

Roly-Polies

with Myra Pasquier

Lesson: Observing and comparing roly-polies to other animals

Learning Objectives: Children will use observation skills to describe the physical features of roly-polies. They will make comparisons with other animals to identify similarities and differences to infer which animal roly-polies are most like.

Materials: Roly-polies, magnifiers, writing and drawing materials, clear plastic plates or tubs, ants and pictures and diagrams of ants, crayfish and pictures and diagrams of crayfish. [Note: Crayfish are most commonly available in the spring; if they are unavailable at the time of your lesson, you can use shrimp instead.]

Safety: Remind children not to put their fingers in their mouth or nose while handling roly-polies or other animals. Make sure children thoroughly wash their hands with soap and water before and after the activity. Students with allergies or immune-system illnesses should not handle the roly-polies or the other animals used in this lesson. Make sure any spilled water is wiped up to prevent slipping hazards. Remind children to keep away from electrical outlets when working with water.

Teacher Content Background: Roly-polies, pill bugs, potato bugs, and doodle bugs are all names for the animal scientists know as *Armadillidium vulgare* (although in different regions of the United States these common names are often given to different animals). We'll call them roly-polies; partly because it's more fun to say but also because bugs are insects and, contrary to popular belief, roly-polies are not insects. They belong to the group of animals called crustaceans and are relatives of crabs, shrimp, and lobsters. Like their crustacean relatives, they have many legs and a body enclosed in a hard, segmented exoskeleton. However, unlike most of their crustacean relatives, roly-polies live on land and are found in almost all areas of the United States, except for deserts—which makes sense, as roly-polies have gills and breathe oxygen dissolved in water and therefore prefer moist environments. The females carry their young in a little pouch under their bellies until the more than 100 young roly-polies are ready to fend for themselves. Roly-polies play a very important role in soil ecosystems, both because they are a food source for many insects and because they are detritivores (they eat, and therefore help to break down, dead and decaying material). Of course, roly-polies are best known for their characteristic defense mechanism of

Searching for roly-polies

curling up into tight little balls, a trick that their relative the sow bug cannot accomplish.

Science terms that may be helpful for teachers to know during this lesson include *observe*, *compare*, (body) *segments*, *antennae*, *legs*, and *environment*.

Procedure

Getting Started

Introduction: Before bringing out the roly-polies, have a discussion with children about being gentle with animals and being sure to not hurt them. Show the children how to handle the roly-polies—demonstrate holding your hand flat and letting a roly-poly uncurl and walk freely across your hand. For children who would prefer not to hold the roly-poly, provide a roly-poly on a clear plastic plate (a clear plate will allow children to see the underside of the roly-poly). Ask children to observe their roly-poly and to describe what it looks and feels like. Encourage children to use a magnifier to get a closer look. Prompt them to notice different parts of the roly-poly body. Ask, "What color is it? Is the top of it the same color as

Observing roly-polies

underneath? How do you think it moves? How many legs does it have? Which end is the front? How can you tell?" And ask the children if they have seen these animals before and if they know what they are called. [Note: Remember that roly-polies have several different names, each of which is acceptable. If most of your children already call them potato bugs or pill bugs or doodle bugs, use that name throughout the lesson.]

Investigating

Observing and Documenting: After children have had an opportunity to observe freely, provide some structure to their observations. Ask children to draw their roly-polies and guide them to include as many details as possible, emphasizing the body segments, legs, and antennae. (Alternatively, you can have children take photographs of their roly-polies. Print these for them, and encourage them to circle or label different body parts on their photograph.) Have children pay attention to the roly-poly's segmented body structure and numerous legs. Ask children to point to what they think might be the head of the roly-poly (the head has two antennae and is usually in front as the roly-poly moves). Sum up children's observations of roly-polies by asking them to count and decide how many body segments the roly-polies have. Do the same for the number of legs. After children have observed, it is helpful if you make a (larger) drawing of your own. Let children instruct you as you draw their observations for them and use questioning to encourage children to provide you detailed instructions (e.g., "What did you observe about the roly-poly? Oh, you saw a head? What shape was it? When I draw an oval for the head, is this how I should draw it? Did your roly-poly look like this?") This type of teacher demonstration will sharpen children's observation skills and help them to create detailed drawings of their own.

Comparing: Bring out the crayfish and ask the children to identify any differences and similarities they might notice between the crayfish and the roly-polies. Share real crayfish (from the seafood section of your local market or, even better, *live* crayfish from a bait shop), as well as downloaded pictures and drawings or diagrams of the crayfish. As simplified representations of the animals, drawings and diagrams will help children to see similarities and differences as they compare. After some initial time for observing this new animal, direct children to analyze their crayfish, focusing on determining the number of body segments and the number of legs. As children direct you, create a drawing of a crayfish that shows the body segments and legs, or have children create a drawing of their own. Repeat this process for ants, drawing three body segments and six legs. By the end of the investigation, children should have drawings of roly-polies, crayfish, and ants to analyze.

Making Sense

Describing Findings: Revisit the drawings and focus the children on the body segments of the animals. Ask them to count the number of parts or segments that make up each animal's body. Support your children as they count. Encourage them to record this information on their drawings and help them to write "7" or "seven segments." Next, direct children toward the animals' legs, asking, "What do you notice about their legs? How many does it have? How many do you see in each part of the body?" Again, support them as they count and record the number of legs.

Generating Explanations: It may be wise to start by helping children to understand the concept of *similar*, explaining that when two things are almost the same, but not exactly the same, we call them similar. Referencing drawings, diagrams, or pictures of the three animals, ask the children which two animals have similar numbers of body segments and which one animal is different. Repeat this comparison for the legs. Explain to your children that "scientists analyze animals the way that we did to find out which animals are related. Animals that have many similarities are often related; these roly-polies and the crayfish have a lot in common. Do you think they're related? Why do you think they are related?" [Note: If you choose to, you can now inform children that roly-polies are not bugs or insects but instead are a type of animal called crustaceans. But much more important than this factoid is the idea that children, just like scientists, can observe animals and find out which ones are similar and which are different.]

Application: Engage children in a discussion of other animals they know that they think are similar and might be related. "What characteristics do these two animals have in common? How else are they similar? Are there any other animals that are similar?" Tell children that it's not just looking the same that determines if animals are related; it can also be sounding similar or acting similar. "Can you think of two animals that sound similar? Can you think of two animals that act similar—that eat similar foods or that live in similar places?" Download pictures of animals to support this discussion; use pairs of closely-related animals, such as zebra–horse, coyote–wolf, kangaroo–wallaby, llama–vicuña, and ostrich–rhea. Be sure to label the pictures so children can see that the names of the animals are different.

What's Next?

Extension Activity

In this activity, children will investigate whether the roly-polies have a preference for wet or dry environments. Ask children, "How can we test to see whether roly-polies prefer wet or dry environments?" Encourage children to ask questions and to explain possible ways to conduct this test. Use a clear tub roughly the size of a shoebox. Have the children place their roly-polies in the center of the tub; there should be about 10 to 20 roly-polies in the tub to make this investigation reliable. Students can wet a napkin (a paper towel or cloth will also work), wring the excess water from it, and then place the wet napkin on one side of the tub; on the other side children can place a dry napkin. The bottom of the tub, and the roly-polies, should be completely covered. Show the tub to the children and let them tell you what is different about the two sides of the tub. Discuss the experiment, asking children to explain how it's set up, what it's designed to investigate, and what results they anticipate. After 15 to 20 minutes, hold the tub up so that children can view the roly-polies from below. Have children describe the movement and placement of the roly-polies they observe. Students can record their data, drawing the roly-polies found on each side (see Figure 2.1, Roly-Poly Data Sheet). Based on the results of the investigation, children can determine if their roly-polies have a preference for wet or dry environments.

Figure 2.1

Roly-Poly Data Sheet

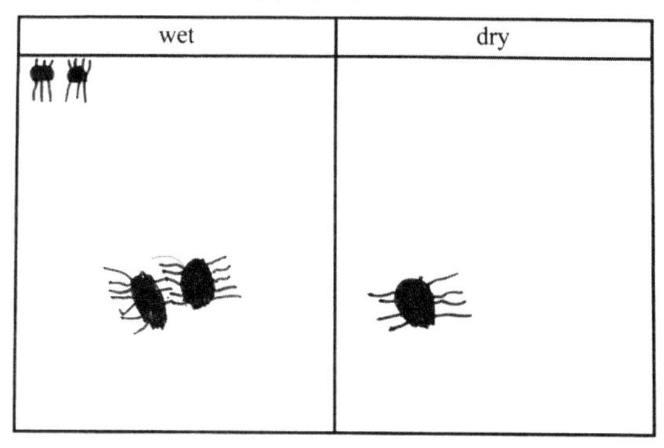

A full-size version of this figure is available at *www.nsta.org/startlifesci*.

Similar experiments can be conducted to investigate additional roly-poly preferences (light vs. dark, different types of food, etc.); let your children decide which variable to test. After each experiment, have children verbalize or write the answer to their research question (e.g., "Roly-polies prefer wet environments"). You can find variations of this experiment described in *Observing Earthworms*, page 30.

Integration to Other Content Areas

Reading Connections

The activities in this lesson include opportunities to draw children's attention to print and to use print in meaningful ways that support children's development of emergent literacy skills. These include verbally describing and drawing roly-polies, making comparisons with different animals, and labeling drawings. Many books, including *Next Time You See a Pill Bug* by Emily Morgan (2013), *A Pill Bug's Life* by John

Himmelman (2000), and *I'm a Pill Bug* by Yukihisa Tokuda (2006), provide informational text paired with images of roly-polies. Students can explore sets of related animals with books such as *The Beetle Book* by Steve Jenkins (2012), *Sorting Snakes* by Mary Rose McDonnell (2013), and *Animals Two by Two* by Lawrence F. Lowery (2015). And the fun book, *Roly-Polies,* by Monica Carretero (2011), informs children about roly-polies while helping children to learn to judge people (or roly-polies) based on their abilities and creativity and not by their appearance.

Writing Connections

With conspicuous *R*, *P*, and *L* sounds, along with long *O* (/ō/) and long *E* (/ē/) sounds, *roly-poly* is a great word for children learning to write letters associated with different letter sounds. Ask your children which letter sounds they hear and help them to write *roly-poly* in a way that makes sense for them. (Expect and applaud, "R O L E" and "P O L E.") With two long *O* (ō) sounds and a body, when curled-up, that resembles the letter *O*, let roly-polies inspire you and your children to search for other things in the class that have the long *O* (ō) sound. For each object found, write a label for the object and have children identify the *O* in the word you have written. Read the word together before going on to the next object.

Math Connections

"Roly-Poly Races" can provide children with several opportunities to use their emergent math skills. Draw a small circle in the center of a large piece of paper and let children place their roly-polies in the circle. Use a timer and encourage children to count with you. (Depending on how active your roly-polies are, about 20 to 30 seconds should be enough time for the race.) At the end of the race, let children mark their roly-poly's position on the paper and measure how far their roly-poly traveled using standard or nonstandard units (paper clips, dominos, and so on). [Note: Although frightened crickets and frogs will jump farther in similar races, the strategies of touching the animal or tapping the ground next to it won't work with roly-polies. Encourage children to be still. Calm roly-polies will crawl farther; startled ones just curl up into a ball.]

Other Connections

Child's Life Connection

Students will be amazed to find how prevalent roly-polies are. Conduct a roly-poly search with your children. Let children know that roly-polies like moist, dark places and invite the children to search in the grass, at the base of bushes, and under rocks. Be sure you've scouted the area ahead of time and have made sure it's free of hazards. If you and your children can't seem to find any roly-polies, try to lure them to you. Place half of a raw potato, cut side down, in a moist grass or dirt area and wait. In the next day or two, you should have roly-polies—check first thing in the morning, but use caution, as roly-polies may not be the only animals beneath the potato. [Note: This technique works well for capturing roly-polies to use in the lesson as well.]

Center Connections

Very few animals are round. Celebrate the roly-polies' ability to curl up into a ball in your centers. In the **art center,** try marble painting. Tape paper (to be painted on) to the bottom of a shallow box. Set out small cups of paint (or use an empty egg carton). Have children place a marble in a paint cup and swirl the marble around to make sure it gets covered in paint. Students can use a spoon to transfer one or more marbles from the paint to the box. Students can then paint by tilting the box and having the marbles roll over the paper, leaving paint trails behind. Ask how the marbles are moving and help children to use the word *rolling*. Fill the **sensory table** (or a tray) with polished stones and add a few marbles. Students can feel among the smooth stones and try to identify the marbles without looking. Encourage children to describe what "round" feels like. For **dramatic play**, have children become "rolling experts." Provide a small ramp and a variety of different objects, along with paper and pencils for children to record their data. Students can test objects to identify those that roll and those that do not. Supply measuring tapes so children can try to determine which objects roll the farthest; encourage children to explain their findings.

Family Activities

In school, your child has been learning about roly-polies. We conducted several different investigations that your child can tell you about and repeat at home. Conduct a roly-poly search with your child. Look in moist, dark places outside. If you can't find any roly-polies, try to lure them to you. Place half of a raw potato, cut side down, in a moist grass or dirt area and wait. Check for roly-polies with your child each morning. Count them together. Let your child tell you about roly-polies and their legs and body segments. Collect a few roly-polies and have your child show you how to test whether roly-polies prefer wet or dry environments. With your child, place roly-polies in the center of a tub and place dry napkins on one side of the tub and wet napkins on the other. Let your child lead. After a minute or two, uncover the roly-polies to see which side of the tub they prefer. Be amazed as your child shows you the results of the experiment and ask your child to explain it to you.

Actividades Familiares

En la escuela, a su hijo se le ha enseñado sobre las cochinillas. Llevamos a cabo numerosas investigaciones sobre las cuales su hijo le puede hablar y repetir en casa. Realice una búsqueda de cochinillas con su hijo. Busque en la humedad, lugares oscuros al exterior. Si no puede encontrar ninguna cochinilla, intente atraerlas. Ponga la mitad de una papa cruda sobre el césped húmedo o en una zona con lodo, con el lado cortado hacia abajo, y espere. Revise si hay cochinillas con suhijo cada mañana. Cuéntenlas juntos. Deje que su hijo que le hable sobre las cochinillas y sobre sus patas y segmentos corporales. Recoja unas cuantas cochinillas y haga que su hijo le muestre como probar si las cochinillas prefieren los ambientes secos o húmedos. Con su hijo, ubique cochinillas en el centro de una bañera y ponga servilletas secas en un lado de la tina y servilletas húmedas en el otro. Deje que su hijo guíe. Luegode uno o dos minutos, descubra las cochinillas para ver qué lado de la tina prefieren. Sorpréndase cuando su hijo le muestre los resultados del experimento y pídale que se lo explique.

Assessment—What to Look For

- ***Can children describe features of roly-polies based on their observations?***
  (using senses to gather information, identifying properties, using new or complex vocabulary)
- ***Can children provide verbal comparisons of different crustaceans?***
  (comparing properties, explaining based on evidence, using new or complex vocabulary)
- ***Can children identify objects as similar and different?***
  (comparing properties, explaining based on evidence, using complex patterns of speech)
- ***Can children record (draw or write) observations and contribute to class discussion?***
  (using new or complex vocabulary, documenting and reporting findings, discussing scientific concepts, listening to and understanding speech)

Standards

Head Start Early Learning Outcomes Framework
P-SCI 4. Child asks a question, gathers information, and makes predictions. *"Asks more complex questions. Uses other sources besides adults to gather information, such as books or other experts. Uses background knowledge and experiences to make predictions."*
Next Generation Science Standards
Science and Engineering Practice: Constructing explanations and designing solutions. *"Make observations (firsthand or from media) to construct an evidence-based account for natural phenomena."*
Common Core State Standards for Mathematics
K.MD.A.2. Describe and compare measurable attributes. *"Directly compare two objects with a measurable attribute in common, to see which object has 'more of'/'less of' the attribute, and describe the difference."*
Common Core State Standards for English Language Arts
SL.1.4/5. Presentation of knowledge and ideas. *"Describe people, places, things, and events with relevant details, expressing ideas and feelings clearly. Add drawings or other visual displays to descriptions when appropriate to clarify ideas, thoughts, and feelings."*

Jumping Crickets

with Nicole Hawke

Lesson: Observing the external features of crickets

Learning Objectives: Children will use vocabulary and drawings to represent detailed observations of crickets and use these observations as a basis for making inferences about the function of crickets' external parts.

Materials: Magnifiers, writing and drawing materials, several crickets each in its own clear container (preferably one large enough to allow crickets to move freely), and small pieces of moist bread (one per container, to feed each cricket)

Safety: Remind children not to put their fingers in their mouth or nose while handling the crickets and make sure they thoroughly wash their hands with soap and water after the activity. Children with allergies or immune-system illnesses should not handle the crickets used in this lesson. Make sure any spilled water is wiped up to prevent slipping hazards. Remind children to keep away from electrical outlets when working with water. Remind children not to eat any food used in this activity.

Teacher Content Background: Crickets are small (usually less than 2 inches long), nocturnal insects found worldwide, with more than 900 different species. Crickets are omnivorous, eating both plant matter and other insects. Their habitats vary widely depending on the species, including fields, forests, caves, and even underground. As with their habitats, crickets vary in appearance as well, occurring in a variety of colors, including shades of black, red, brown, and green. Conspicuous on a cricket's body are two long antennae, sometimes called feelers, that help crickets to find food; two wings, although crickets do not fly; and three pairs of legs with the hind pair larger and longer than the others. Most people identify crickets by the loud chirping sound they produce; however, only the male crickets chirp, producing this "song" by rubbing their wings together. The easiest way to determine the sex of most common crickets is to look for the long, straight ovipositor extending from the rear of the cricket. Found only in females, the ovipositor functions to deposit eggs. Unlike many other insects, crickets do not undergo a complete metamorphosis; instead, young crickets look just like smaller versions of the adults.

Science terms that may be helpful for teachers to know during this lesson include *investigate*, *observe*, *antennae*, *legs*, *compound eye*, *wings*, *cerci*, and *ovipositor*.

Procedure

[Note: To help ensure the safety of the crickets, use only a few crickets and keep them all securely contained throughout this lesson.]

Getting Started

Introduction: Present children with a variety of pictures of crickets. Share pictures of crickets up close and also of crickets in their habitat. Prompt children to look carefully at the pictures and call on children to share with the group what they see in the pictures. Ask the children if they know the name of the insect in the picture and clarify for the children that this insect is called a cricket. Then prompt children to look closely at the cricket's body and encourage children to identify the different body parts they observe (e.g., legs, antenna, head).

Child Questions: Invite the children to think of and ask any questions they have about crickets. At this time, children may be excited to share what they already know about crickets and the experiences they have had with crickets. Be sure to foster this enthusiasm, but at the same time, encourage children to share what they wonder about crickets. Record these questions in a visible place and explain to the children that as they investigate crickets they may discover answers to some of their questions.

Initial Explanation: Refer children back to the pictures of the crickets and ask children how they think crickets move. [Note: If you want a more challenging lesson, focus children's thinking on how crickets sense rather than on how they move.] Prompt children to notice any clues in the pictures that might suggest how crickets move. Give each child a picture or diagram of a cricket and ask children to identify parts of the cricket that help the cricket to move. Students can describe to a classmate how they think crickets move. If children have difficulty thinking about motion, ask them if they think crickets "slither like a snake? swing like a monkey? hop like a bunny? swim like a fish?" and ask children to think about the body parts necessary for these types of motion. As children describe and even act out their ideas, there may be some very enthusiastic demonstrations of cricket motion—encourage this enthusiasm.

Investigating

Observing: Before distributing crickets for children to observe, reassure children that crickets do not bite or sting and inform children that for their observations they will be using only their sense of sight. Remind them how to use a magnifier, and have a few children model this for the class. For the first few minutes, have children simply observe the crickets. While they make their observations, listen to the comments you hear children making and, whenever possible, follow their leads. Ask children to describe their observations to you. Use questioning to get children to expand their descriptions and then to direct their attention to the external parts of the cricket (e.g., the eyes, the back legs, the antenna) and to how their cricket moves. As children look at the external features on the crickets, encourage them to count. This will help focus children's observations and help children to develop one-to-one correspondence as they count and learn more about the different body parts of crickets (e.g., six legs, two antennae).

Documenting: After these initial observations, ask the children to record their observations by making a drawing of the cricket. Provide time for children to add as much detail to their drawing as they can. As they draw, encourage children to add detail and to explain what their drawing shows; ask questions such as "Does your cricket have eyes? Where are the eyes? Is there anything else on your cricket's head? Do the antennae move? Can your cricket move? How does your cricket move? Do all the legs help the cricket jump? How do the different legs help the cricket to move?" (If you'd prefer, instead of drawing the cricket, you can provide children with a picture of a cricket. Students can identify and label the different external parts on their cricket picture.)

Crickets for observation

Making Sense

Describing Findings: Display the children's drawings and have children take turns sharing their drawings with the class, developing their oral language skills as they describe aloud what their drawing shows. When the children are done presenting, ask them to recall all the different cricket body parts included in the drawings. With your children's guidance, create a class drawing of a cricket and label the external features that the children have described. Possible features that children might notice include legs, body, head, eyes, antennae, cerci, ovipositor, and possibly wings.

Generating Explanations: Using a drawing or diagram of the cricket's external parts, challenge the children to think about what each body part is used for. Here it may be useful to have the children observe the crickets again, one part at a time. Have the children return to their drawings and look to see if they included each of the external parts discussed. Have children practice pointing at one of the external structures and explaining how the cricket uses that part. For example, children may point to their cricket's hind legs and say, "The cricket uses its legs to jump." Throughout this discussion, keep in mind that the "right" answer is not the goal. Engage children in the process of generating explanations (right or wrong) based on their observations.

Curiosity: Now is a good time to refer back to the list of questions children had about crickets. Most likely, some of the questions that were posed have been answered through their observations, but others may not have been. Review each question and ask children to share what they've learned during the lesson as they offer answers to the questions. Then ask children if their observations have brought about any new questions about crickets. Help children to act on their curiosity—plan further investigations with the crickets and share books or videos that may help address children's questions and might lead to even more questions and new explorations.

What's Next?

Extension Activity

After looking at the external structures of the cricket, choose another animal to compare; select an animal that can be brought into the class for observations. (Choosing something like a snail or an ant will provide the children with something that is quite a bit different but also has several similarities to be discovered.) Comparing the animal and the cricket will help children to see that animals have external structures adapted to fit their habitat and needs. Just as you did with the cricket, complete a drawing of the new animal illustrating its external parts and discuss what each may be used for. Have the children observe the animal and draw it. Once this is done, lead the class in a discussion on what external parts they saw on the new animal. Then have children go back and observe one structure at a time to try to identify what it is used for. Finally, complete the class drawing that includes labels of the structures and brief descriptions of their use for the new animal (e.g., legs—walk, eyes—see). Now your children can compare crickets and the new animal and discuss which body parts are the same and which are different. Students can consider why these parts would be different or the same as you ask them to compare how the two animals sense things around them, how they move, and how they protect themselves.

Integration to Other Content Areas

Reading Connections

Throughout this lesson, there are ample opportunities to develop children's literacy skills and to engage them with print. From the list of class questions about crickets to cricket drawings and labeling, children are seeing the importance of print and the message it carries.

To further this understanding, share with children many different trade books related to crickets. Students can learn more about crickets while experiencing the features of informational text through the use of nonfiction books about crickets and their body parts, life cycle, and singing, such as *Chirping Crickets* by Melvin Berger (1998), *Crickets* by Cheryl Coughlan (1998), and *Singing Crickets* by Linda Glaser (2009). Also, through reading quality literature about crickets, such as *The Very Quiet Cricket* by Eric Carle (1997) and *A Pocketful of Cricket* by Rebecca Caudill (2004), children can see that information in science can also be used in creative and entertaining ways when we use our imaginations.

Writing Connections

During this science investigation, children have several opportunities to further develop their emergent literary and language skills, such as when children are making and recording their cricket observations and when children describe these observations aloud. This connection between writing and science can be extended by helping children to add labels to their observations. Ask children to revisit their observational drawings and encourage children to add some of words the class has been learning to their observations. If needed, write for children the labels they suggest or provide labels for children to glue onto their drawings. Through this experience of labeling their observations, children receive the message that print contains meaning. Also, by creating, labeling, and sharing their observations with the class, children can understand that scientists use drawings and written words as tools to share what they have learned about our world.

Math Connections

Outside, children can try to measure cricket jumps. Place a cricket on the sidewalk or other paved surface. Have children use chalk to mark the place where a cricket begins and ends a jump. Students can then measure the distance between marks using string, or rulers, or other measuring units (e.g., blocks, paper clips, toy crickets). Students can measure several different jumps to find out which was the longest. [Note: Crickets generally crawl and use hopping as a way of escaping predators. Students can coax a cricket into jumping by touching it, but keep in mind that to the cricket, each hop represents a near-death experience. Return the cricket to safety after a few hops. Also, be sure that your crickets do not escape, as store-bought crickets are likely to be a nonnative species; their release can cause harm to native species.] Children can pretend to be crickets and measure their own jumps or count how many jumps they take to move from one place to another.

Other Connections

Child's Life Connections

Have children share with you their experiences with crickets. Ask children where they have seen crickets and to name places they might find crickets (campground, pet store, backyard, and so on.). Tell children that crickets sing a song and ask if any of the children have heard crickets before—maybe when they were outside on a summer night. Invite children to mimic their ideas of the cricket song. Then play a recording of the cricket song. Students may recognize the sound, but may not have known that it was a cricket using its wings to sing! Let children share when and where they've heard crickets singing. Challenge children to use their voices to mimic the cricket's song and encourage children to listen for crickets at their homes just before bedtime.

Center Connections

At the **art center**, provide children with supplies to paint a picture of a place they might see a cricket (in grass, under rocks, etc.). Then provide children with a variety of pictures of crickets and ask each child to select one picture and to cut out the cricket in the picture. Once their painting is dry, children can glue their crickets onto the habitats. At the **sensory table**, add toy crickets. Students can imitate how crickets move with the toys. They can also act out other cricket behaviors they may have observed, such as eating or moving their antennae. As children play with the toy crickets, encourage children to use some of their new academic language to describe their cricket and how it's moving. For **dramatic play**, consider setting out musical instruments. Include a variety of instruments that involve rubbing action such as sand paper blocks, guiro, and rhythm sticks; playing these instruments can help to promote children's motor development. Encourage children to create and perform their own cricket songs. Ask the children to share the meaning of their songs with you.

Family Activities

Your child has been learning about different animals. This week we studied crickets. You can promote your child's learning by asking him or her to describe a cricket to you and to share with you what they know about crickets. Show your interest in learning by asking your child questions about crickets such as, "How many legs do crickets have? How far can crickets jump? What do crickets sound like?" As a family, go out one evening to search for crickets. Give your child a small net or paper cup and see if he or she is able to catch any crickets. Have your child listen carefully for the song of the cricket. Let your ears guide you on your cricket search. If you catch a cricket, let your child point out the different parts of the cricket's body before you release it. Be an enthusiastic learner as your child teaches you about crickets.

Actividades Familiares

Su hijo ha estado aprendiendo sobre distintos animales. Esta semana estudiamos los grillos. Puede promoversu aprendizaje pidiéndole que le describa al grillo y que le diga lo que sabe sobre estos insectos. Muestre interés en aprender haciéndole preguntas sobre grillos, como: "¿Cuántas patas tienen los grillos?" "¿Que tan lejos pueden brincar los grillos? ¿Cómo suenan los grillos? Como familia, vayan una tarde en busca de grillos. Entréguele una pequeña red o un vaso de papel y vea si puede atrapar algún grillo. Dígale a su hijo que debe escuchar el sonido del grillo atentamente. Deje que sus oídos lo guíen en su búsqueda de grillos. Si atrapa un grillo, indíquele a su hijo que le muestre las distintas partes del cuerpo del grillo antes de liberarlo. Muestre una actitud entusiasta por aprender cuando su hijo le enseñe sobre los grillos.

Assessment—What to Look For

- **Can children describe crickets based on their observations?**
  (using senses to gather information, identifying properties)
- **Can children compare their initial and observational drawings of crickets?**
  (comparing properties, explaining based on evidence, using new or complex vocabulary)
- **Can children infer the purpose of body structures on the crickets?**
  (constructing explanations, explaining based on evidence, making inferences, using complex patterns of speech)

Standards

Head Start Early Learning Outcomes Framework
P-SCI 2. Child engages in scientific talk. "Uses scientific content words when investigating and describing observable phenomena, such as parts of a plant, animal, or object."
Next Generation Science Standards
Science and Engineering Practice: Constructing explanations and designing solutions. "Make observations (firsthand or from media) to construct an evidence-based account for natural phenomena."
Common Core State Standards for Mathematics
1.MD.A.2. Measure lengths indirectly and by iterating length units. "Express the length of an object as a whole number of length units, by laying multiple copies of a shorter object (the length unit) end to end; understand that the length measurement of an object is the number of same-size length units that span it with no gaps or overlaps."
Common Core State Standards for English Language Arts
W.K.2. Text types and purposes. "Use a combination of drawing, dictating, and writing to compose informative/explanatory texts in which they name what they are writing about and supply some information about the topic."

Observing Earthworms

with Lauren M. Shea

Lesson: Observing the external features and behavior of earthworms

Learning Goals: Children will observe and describe characteristics of earthworms and compare them to other animals. Students will investigate how various conditions affect the earthworm.

Materials: Various photographs of earthworms (e.g., showing worms eating, in the rain, in the soil), live earthworms (nightcrawlers or red wigglers), writing and drawing materials, magnifiers, clear plastic cups, moist paper towels, plastic spoons, small bowls of water, soil

Safety: Remind children not to put their fingers in their mouth or nose while handling earthworms or other animals. Make sure children thoroughly wash their hands with soap and water before and after the activity. Children with allergies or immune-system illnesses should not handle the earthworms or the other animals used in this lesson. Make sure any spilled water is wiped up to prevent slipping hazards. Remind children to keep away from electrical outlets when working with water.

Teacher Content Background: As a primary contributor to enriching and improving soil for plants, animals, and even humans, earthworms (*Lumbricus terrestris*) are an important component of soil ecosystems. Earthworms burrow into the soil, creating tunnels that aerate the soil, allowing air, water, and nutrients to reach deep within the soil and closer to plant roots. Earthworms can eat their way through the soil, digesting organic material (mostly decaying plant material) as they go. The "castings" that these decomposers excrete contain phosphorus, nitrogen, and other important nutrients, making soil ecosystems more rich and productive.

Earthworms' bodies are made up of 100 to 150 ring-like segments called annuli. These segments contain muscles and have small bristles (called setae) extending from their surface. The setae can be extended or relaxed; when extended, they serve as tiny anchors grabbing the soil. Earthworms move and burrow into the soil by alternately contracting and relaxing their muscles and setae, causing their bodies to lengthen in one area or contract in other areas. The typical earthworm is about 3 inches in length, but some have been known to grow up to 14 inches long. Earthworms do not have lungs; they take in oxygen from the air through their skin. In order for an earthworm to breathe, its skin must be kept moist. Earthworms therefore live in damp or moist soil. (The ones you collect will require occasional misting with water throughout the lesson.) Earthworms can also be found on the surface of the soil (or sidewalk) when it rains; these wet conditions allow them to travel overland without drying out.

Science terms that may be helpful for teachers to know during this lesson include *observe, experiment, characteristic,* (body) *segments, clitellum,* and *setae.*

Procedure

Getting Started

Introduction: To engage children's interest in earthworms, read aloud *Diary of a Worm* by Doreen Cronin (2003) and encourage children to pay particular attention to the appearance, characteristics, and behaviors of the earthworm in the story. While reading aloud, model how you ask yourself questions as you read. For example: "Hmm, I wonder, do earthworms really live in the soil? What do earthworms really eat? Do earthworms really do homework?" Welcome children to share questions of their own. Write out their questions on chart paper, using picture support as appropriate.

Curiosity: After reading and discussing questions about the book, show children several photographs of real earthworms and encourage children to come up with more questions about earthworms. If needed, prompt children using questions, such as "Where can earthworms be found? What do you notice about the earthworm's body?" Refer to the children's questions throughout the remainder of the lesson and point out when children's explanations provide insight to their questions as you learn together.

Investigating

Observing: After a rainy day, take your children outdoors to search for earthworms. Give each child a paper cup and a spoon and remind them to show care as they scoop up earthworms. Ask children for ideas about where to search for the earthworms and direct them to appropriate areas. [Note: If your school yard isn't particularly "wormy," you can purchase worms online or at your local bait shop and, the day before the lesson, plant them in loose soil in areas children will explore.] As the children dig, engage them in conversations about the earthworms they find; prompt children to share where and how they found the earthworm, asking questions such as, "Did you find the earthworm close to the surface? What did the earthworm do when you scooped it up? Did you notice how the earthworm was moving in the soil? What do you think the earthworm was doing when you found it?" Bring the collected earthworms back to the classroom and give children magnifiers to observe earthworms more closely. Place the earthworms on a flat surface and allow children to touch or hold the earthworms if they choose to. Be sure to discuss the importance of being respectful and gentle with living things (see the information at the beginning of the chapter). Remind children to use their sense of sight and touch to make detailed observations of the earthworms. Listen to children as they share their discoveries and guide your questioning around what seems to interest them. When appropriate, guide children to consider earthworms' body parts ("Do you notice if earthworms have eyes? Legs? A mouth? What parts can you see?") and behaviors ("How does your earthworm move? What does it feel like? What do you think the earthworm might eat? Why do you think that?"). As children observe, demonstrate your own enthusiasm for earthworms as you rephrase and validate their responses ("Yes, you're correct, I also notice that the earthworm does not have feet. I wonder how it is able to move!"). Throughout this time of exploration, record children's ideas,

Observing an earthworm

thoughts, questions, and answers to discuss later in the lesson.

Comparing: Gather the children around, place three earthworms in front of them, and ask the children to tell you what is the same and what is different about these earthworms, comparing size, color, body parts, and other observable characteristics. Encourage children to discuss similarities; children may notice that all earthworms are wiggly or that they all have lines (segments) or a "puffy" part in their bodies (clitellum). Ask children if the earthworm they observed earlier also had these features. If you've completed other animal investigations, prompt children to compare the earthworms to other animals they have studied.

Making Sense

Explanation Generation: After several opportunities to observe and discuss earthworms, children may be ready for more formal concept development. Ask the children, as a group, to discuss what they have learned about earthworms. Review children's questions from the beginning of the lesson and determine what they have learned about earthworms. Revisit the comparisons children made during the *Investigating* section of the lesson, asking children to explain features that earthworms have in common. Confirm their observations with statements such as "Yes, all earthworms have many segments." Use a large picture or create a drawing of an earthworm to help children point out the structures they notice. If children discussed other animals that they've studied (snails, roly-polies, crickets, and so on), encourage children to describe how earthworms are similar to and different from those animals. Throughout this discussion, children may provide answers to questions asked during the *Getting Started* section of the lesson; record these answers on the chart paper next to the corresponding question.

Curiosity: Ask the children what else they would like to know about earthworms and brainstorm together other investigations that could help you find out more about earthworms together. Nurture and share in their curiosity. Guide children as they continue earthworm investigations that stem from their own interest and wonderings.

What's Next?

Extension Activity

After initial explorations of earthworms, have your children design and conduct an experiment investigating earthworm preferences. Have children brainstorm and discuss different variables that might affect earthworms and let children decide which ones to test. The possible variables to test are endless; consider: Do earthworms prefer soil with leaves added to it or not? Rocks added or not? Pieces of carrots added or not? Sand added or not? Record children's ideas and then together decide which variable to test. [Note: You may decide to limit the options available for this experiment. If so, choose and present a short list of variables to the children. Either way, be sure to let children decide which variable they are interested in testing.] Fill a large, uncovered plastic container with moist soil or planter mixer. [Note: Be sure to keep the soil moist at all times.] Have children add the variable that they've decided to test to the soil in one half of the container and place five worms in the center of the soil. Over the next few days, have children visit the experiment and locate each of the earthworms. Provide a place for children to record their data using numbers or tally marks (see Figure 2.2, p. 34). After children have had a chance to gather and record data, they can find the totals for each half of the container and interpret the results.

Figure 2.2

Worm Data Sheet

Worm Data		
	[with variable]	[without variable]
1		
2		
3		
4		
5		
6		
7		
8		
9		
10		
Total		

A full-size version of this figure is available at *www.nsta.org/startlifesci*.

Integration to Other Content Areas

Reading Connections

This lesson has many opportunities for children to develop their emergent literacy skills as they learn about earthworms. Fill your classroom library with books about earthworms for children to explore. Entertaining fiction options include *Diary of a Worm* by Doreen Cronin (2003); *Wiggle and Waggle* by Caroline Arnold (2009); and *Winnie Finn, Worm Farmer* by Carol Brendler (2009). Nonfiction books such as *Garden Wigglers* by Nancy Loewen (2005), *Wonderful Worms* by Linda Glaser (1994), *An Earthworm's Life* by John Himmelman (2001), *Wiggling Worms at Work* by Wendy Pfeffer (2003), and the narrative text of *Yucky Worms* by Vivian French (2012) will help you and your children learn more about these amazing animals. As you read these books, fiction or nonfiction, encourage children to talk about what they know about earthworms and to share any new questions they have or insights they learned from the books.

Writing Connections

Throughout the lesson, there are opportunities to support children's content development and to reiterate the role of documented communication as children draw and write to express their learning. For additional writing opportunities that will support their continued sharing, learning, and emergent writing experiences, have children help you create an earthworm-themed bulletin board. Ask for children's input on the design and content of the bulletin board; include what they think is important to show about earthworms. Students can draw and write their favorite facts about earthworms, recording details from their observations. Support children's letter formation and sound recognition as they write and, if children are ready for sound–letter correspondence, encourage them to label the parts of the earthworm in their drawing.

Math Connections

During the lesson, children will have observed that earthworms vary in size, providing an excellent opportunity to develop measurement skills. While you have the earthworms, you can model how to measure one using standard or nonstandard units (e.g., "Placing the beginning of my ruler at the beginning of the earthworm, I can measure that it is 3 inches long"). Students can meaure and compare the lengths of different earthworms. [Note: Caution children to not stretch their worms as they measure.] Also, children can use "worms" as nonstandard units of measure. Provide children with pretend earthworms (e.g., toy earthworms, pieces of yarn, gummy worms, linguini or spaghetti) and let them compare their pretend earthworms to other objects in the room. With each, you can have the children state, or write, or draw (with sentence starters): "My pretend earthworm is longer than a block. My pretend earthworm is shorter than a marker." Measuring is an important emergent skill for young scientists, allowing them to begin the processes of comparing, describing attributes, interpreting data, and using tools to identify properties.

Other Connections

Child's Life Connection

At the end of your earthworm investigations and activities, discuss with children what the class should do with the earthworms. Students may want to keep the earthworms as class pets. Discuss with children what earthworms need to survive and talk about children's role in ensuring the basic needs of your pet earthworms, such as keeping them moist and feeding them. (See *Animal Walk*, p. 131, to learn more about animals' basic needs.) If you'd prefer to return the earthworms to their homes, use the wonderful book *The Bog Baby* (by Jeanne Willis 2009) to prompt a discussion about the needs of earthworms and why it's important to return them to their natural home.

Center Connections

Learning about earthworms allows for many activities in your classroom centers. At the **art center**, have children make worm paintings. Students can dip thick pieces of yarn or cooked linguini or spaghetti in paint and wiggle these "paint brushes" around the paper, making "worm tracks" on their paper. At the **sensory table**, give the children brownish playdough. They can roll the playdough to create various-sized earthworms. The longest earthworm ever recorded was 40 centimeters (just over 15.5 inches) long. Show your children a model of this world-record earthworm and challenge your children to roll the longest playdough earthworm that they can. For **dramatic play**, children can lie on their bellies and pretend to be worms navigating an obstacle course. Using the playground or classroom, children can set up different courses for each other. They will develope their gross motor skills as they wiggle through tunnels, go under chairs, move around buckets, and dig deep down a slide.

Family Activities

This week in school, your child has been studying earthworms. You can ask your child about her or his observations and investigations; she or he will have lots to share with you. You can further support your child's learning by going on an earthworm hunt together. On, or right after, a rainy day, head outdoors to observe water puddles and muddy places in search of earthworms. When you find an earthworm, talk with your child about how earthworms feel, move, and eat. Let your child be the expert, telling you what she or he sees, knows, and thinks about earthworms. Ask questions such as "What do you notice the earthworm is doing? How is it moving? What do you think it eats?" If you have a hard time finding worms, dig around a bit with a spoon or shovel. If your earthworm hunt is unsuccessful, or if you and your child want to learn even more about earthworms, head to your public library for books about earthworms. Here are some of our classroom favorites:

Actividades Familiares

Esta semana en la escuela, su hijo ha estado estudiando las lombrices de tierra. Puede preguntarle sobre sus observaciones e investigaciones; tendrá mucho que compartir con usted. Puede apoyar el aprendizaje de su hijo aún más, llevándolo de paseo a una búsqueda de lombrices juntos. En o inmediatamente después de un día lluvioso, vaya afuera al aire libre para observar charcos y lugares lodosos para buscar lombrices de tierra. Cuando encuentre una lombriz de tierra, converse con su hijo sobre cómo se sienten, mueven y comen las lombrices. Deje que su hijo sea el experto, que le diga lo que ve, sabe y piensa sobre las lombrices. Hágale preguntas como: "¿Qué ves que está haciendo la lombriz? ¿Cómo se está moviendo? ¿Qué crees que come?" Si se le hace difícil encontrar lombrices, cave un poco en los alrededores con una cuchara o pala. Si no tiene éxito con la búsqueda de lombrices o si usted y su hijo desean aprender más sobre las lombrices de tierra, visiten su biblioteca pública para encontrar libros sobre lombrices. Estos son algunos de nuestros favoritos de clase:

[Teacher: insert list of earthworm books used in class here.]

Assessment—What to Look For

- **Can children describe earthworms based on their observations from the real critter's physical attributes, movements, and actions?**
  (using senses to gather information, identifying properties, using new or complex vocabulary)
- **Can children communicate their learning about earthworms?**
  (using new or complex vocabulary, discussing scientific concepts, listening to and understanding speech)
- **Can children recognize commonalities and differences among earthworms and other animals?**
  (comparing properties, explaining based on evidence, using complex patterns of speech)

Standards

Head Start Early Learning Outcomes Framework
P-SCI 1. Child observes and describes observable phenomena (objects, materials, organisms, events, etc.). "Makes increasingly complex observations of objects, materials, organisms, and events. Provides greater detail in descriptions. Represents observable phenomena in more complex ways, such as pictures that include more detail."
Next Generation Science Standards
Science and Engineering Practice: Planning and carrying out investigations. "Make observations (firsthand or from media) and/or measurements to collect data that can be used to make comparisons."
Common Core State Standards for Mathematics
1.MD.A.2. Measure lengths indirectly and by iterating length units. "Express the length of an object as a whole number of length units, by laying multiple copies of a shorter object (the length unit) end to end; understand that the length measurement of an object is the number of same-size length units that span it with no gaps or overlaps."
Common Core State Standards for English Language Arts
SL.K.1. Comprehension and collaboration. "Participate in collaborative conversations with diverse partners about kindergarten topics and texts with peers and adults in small and larger groups."

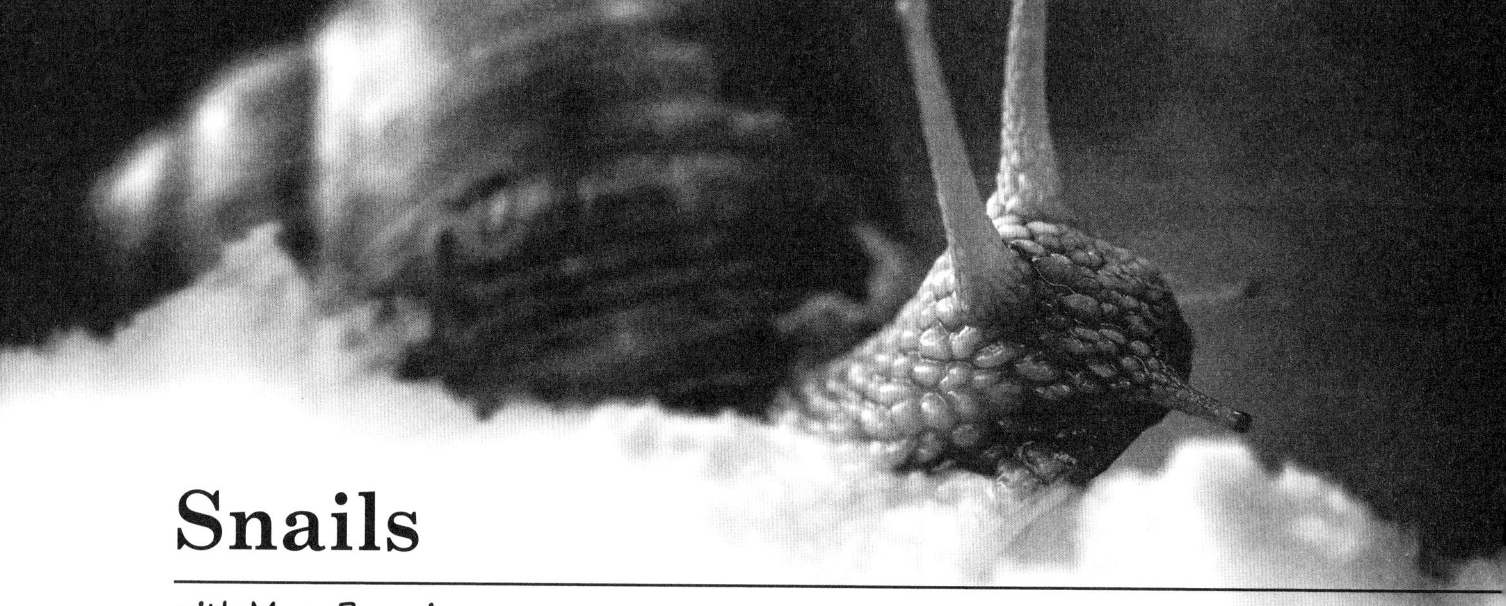

Snails

with Myra Pasquier

Lesson: Observing the external features and behavior of snails

Learning Objectives: Children will use observation skills to describe the characteristics of snails. Students will infer the role of the mucus or slime in snail movement and compare the amounts produced on different surfaces.

Materials: Large brown garden snails, pictures of snails and "snail trails," writing and drawing materials, magnifiers, fine sandpaper or construction paper, acetate sheets, transparent containers or trays

Safety: Use only *Helix aspersa* for this activity, as it is not known to carry any organisms that may be harmful to children. (Other species of snails are known to host harmful parasites.) Be sure to instruct children to not put their fingers in their mouth or nose while handling the snails, and make sure children thoroughly wash their hands with soap and water before and after the activity. Also, due to agricultural and environmental concerns, this snail is currently on the U.S. Department of Agriculture (USDA) list of plant pests; snails purchased for this lesson should not be released outside. Snails can cause significant damage to gardens, crops, and native species. Children with allergies or immune-system illnesses should not handle the snails or the other animals used in this lesson. Make sure any spilled water is wiped up to prevent slipping hazards. Remind children to keep away from electrical outlets when working with water. Use caution in handling sandpaper as it can scrape and cut skin.

Teacher Content Background: The brown garden snail (*Helix aspersa*) is mainly active at night, but will usually come out during the day after it rains. These snails prefer moist habitats and tend to live in clusters under rocks and other structures. They can feed on a wide variety of plant matter, allowing them to survive and thrive in various locations. Snails move by using a gliding or sliding motion, propelled by a long, flat, muscular organ called a foot. Snails secrete mucus to help their foot glide over rough surfaces. Snail shells vary in color (ranging from dark brown to pale yellow), pattern (some with dark, bold bands; others with nearly inconspicuous bands), and size (from less than a half inch in diameter for the youngest snails to nearly 2 inches). *Helix* snails are native to Europe and were introduced to North America, possibly as a food source. Due to their easily adaptable nature, *Helix* snails have become widespread pests to crops and ornamental plants. Garden snails hibernate in the winter, so this lesson is best taught in the spring. And just so you know—snails are not slugs with shells; snails and slugs are two different, although related, animals.

Science terms that may be helpful for teachers to know during this lesson include *observe, experiment, shell, foot,* and *mucus*.

Procedure

Getting Started

Prior Knowledge: To grab children's interest, show them a picture of a trail left by a snail. Ask them if they know what it is or what might have caused it. After children have shared their initial ideas, show them a picture of a snail with a trail. Ask children if they have ever seen any of these animals and if they know what they're called. Probe children's prior knowledge about snails. Find out what they know about snail anatomy and behavior; ask questions such as "What are the different parts of a snail? How do snails move? Where can you find snails? What do snails need to live?"

Curiosity: Ask children what they would like to learn about snails. Generally, the characteristic that children are most interested in is the "slime" that snails leave behind them. Inform children that scientists call this slime *mucus* and support your children's understanding of this new word. Encourage children's interest in mucus. Ask, "Why do snails leave trails of mucus behind them? What do they use the mucus for?" Invite children to offer their speculations and let them know that they'll get a chance to investigate snails and snail mucus.

Investigating 1

Observing and Documenting: Provide children with a magnifying lens and a snail on a transparent tray. Ask them to observe how the snail looks and moves, as well as anything about the snail that they find interesting. You may choose to remind children that scientists record their observations and to encourage them to make scientific observations by looking carefully and drawing what they see. For the youngest learners, consider providing an outline of a snail to support their drawing or have children take photographs of their snails. Print these for children, and encourage them to circle or label specific parts of the snail on the photograph. As they observe and draw, focus children's attention on physical aspects of the snail—have them notice different features of the snail and describe the size, shape, color, and texture of each feature. Your children may notice several snail body parts (e.g., shell, head, tentacles, eyestalks, foot, mouth); if children are recording or drawing their observations, urge them to include these features in their drawings. Then direct children's attention to how the snail moves. Ask, "Can you describe how your snail is moving?" After looking from above, lift the tray for children so they can observe the underside of the snail. You can point out to children that this part of the snail's body—the part that makes contact with the ground as it moves—is called the foot. Ask them to observe and describe the foot as well as the motion of the snail. Make sure the children notice the gliding motion as snails move and the trail of mucus they leave behind. Prompt children to explain or demonstrate how the foot of the snail contracts and expands, propelling the snail forward.

Investigating 2

After children have had an opportunity to observe snails moving, it's time to conduct a snail mucus experiment. To prepare for the experiment, line one half of a tray with a sheet of acetate and the other half with sandpaper or construction paper.

Observing a snail through a magnifying glass

Experimenting: Let the children feel each half of the tray. Ask them to describe what each half feels like and to compare and contrast the two. Before testing the surfaces, ask children what they think might happen when the snail tries to crawl on the acetate and on the paper and encourage children to explain their reasoning. Then have a child place a snail on the acetate-lined half of the tray. Ask children to observe and describe the motion of the snails, noting how the foot of the snail contracts and expands. After the snail has crawled on the acetate, have a child transfer the snail to the paper-lined half of the tray. Students should observe and note any differences in snail movement in this half of the tray. Ask them to make any observations about the mucus left behind on the two surfaces; does it appear that snails produced more mucus on the rough surface or the smooth one? (See Figure 2.3.)

Making Sense

Generating Explanations: The data from the experiment in Investigation 2 should show that snails produce more mucus on the rougher surface. With examples of both the acetate and the paper to refer to, have children explain the data, speculating why snails would produce more mucus on rough surfaces. The goal is not to have children produce or learn the "right" answer, but rather to engage (right or wrong) in the process of making sense of observations and explaining data. Emphasize the experience and ensure that all children have an opportunity to explain. You may want to use a line of questioning to help children to generate explanations that make sense and connect multiple observations. For example, "snail's mucus is slippery. Why do you think snails use more slippery mucus on the rough sandpaper? What would happen to a snail's soft body on the rough sandpaper without mucus? Do snails need slippery mucus more on smooth plastic or rough sandpaper?" But before you provide these questions, give children a chance to make this connection on their own and, more important, emphasize children's ideas over the "right" answer.

What's Next?

Extension Activity 1

One of the most obvious ways that individual snails differ from each other is the size and coloration of their shells. Encourage children to investigate snail shells. Have children compare two different snails (purposely select snails with obvious differences in shell color, pattern, size, and so on.). Ask, "Do you notice any features of these snails that are different?" Lead children to focus on the shell. Have them describe the size and shape of the shell. Tell them to pay attention to differences between the bigger snails and the smaller snails. "Do the shells of bigger snails have the same number of spirals as the smaller snails?" Model for children how to count the spirals. Give children a few different snails and have them sort the snails into two groups based on different shell characteristics. For example, children could sort the snails with large shells from the ones that have small shells, snails with light-colored shells from ones with dark shells, snails with shells that have lines or bands on them from ones that do not have bands, and so on.

Figure 2.3

Snail Experiment

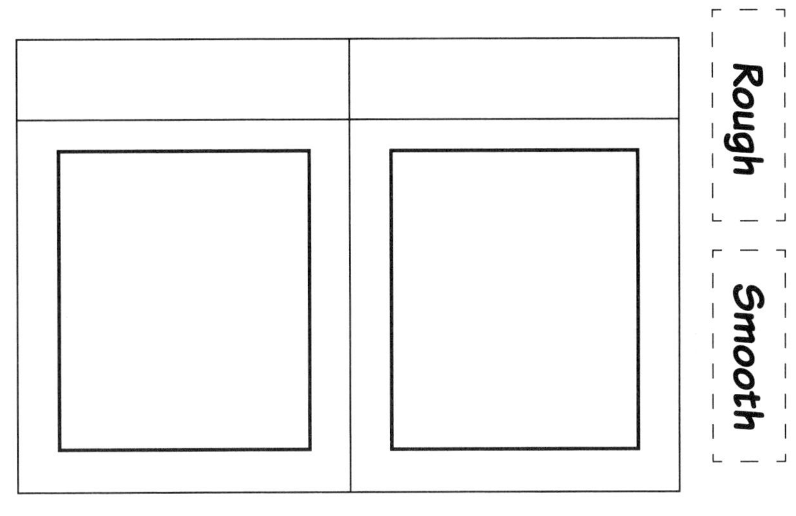

A full-size version of this figure is available at *www.nsta.org/startlifesci.*

Extension Activity 2

Ask older children why they think snails have shells and listen to their responses. Students will commonly say that the shell is the snail's house or home. When they do, lead children to understand the important purpose for snail shells by engaging in a line of questioning such as "How does a house help us when it rains? Yes, our houses keep us dry, but snails like it when it's wet and can't survive when it's hot and dry. How do you think the snail's shell helps it to survive when it's hot and dry? Yes, a snail's shell helps to seal in the moisture that it needs." To demonstrate this idea, you and your children can conduct a shell experiment by placing two containers of water in the sun: one with a "shell" (a lid with a picture of a snail shell) and one without. After a few hours in the sun, compare the two containers; let children observe and describe how the "shell" helped to keep the water in.

Integration to Other Content Areas

Reading Connections

The activities in this lesson include ways to draw children's attention to print and to use print in meaningful ways that support the development of children's emergent literacy skills. These include describing, drawing, recording, and explaining the movement of their snails. There are many snail books to share during reading time. *Snail Trail* by Sally Grindley (2013) and *Slow Snail* (2013) by Mary Murphy are books for early readers that have snail movement, and the trails they leave behind, at the center of their stories. Students can learn more about snails with books such as *Are You a Snail?* by Judy Allen and Tudor Humphries (2013), which provides information about a snail's life in a narrative format, and with informational texts such as *Let's Look at Snails* by Laura Hamilton Waxman (2009) and *A Snail's Pace* by Allan Fowler (1999).

Writing Connections

Students can write a "lab report" using pictures and, as appropriate, words to describe the experiment from Investigating 2. Title children's reports "My Snail Experiment" and allow children to draw or dictate descriptions of how the experiment was set up and the results that were found. You can provide small cutouts of used acetate and sandpaper or construction paper along with labels with the words *smooth* and *rough* for children to glue on each side of their lab report drawing. At the bottom of the report, provide the sentence "Snails use more mucus on the ___ surface" for children to complete (see Figure 2.3, p. 41). This type of writing helps children to understand that writing helps to communicate information and engages children in a writing practice that is very similar to that of scientists.

Math Connections

Students can measure the dried snail trails on the acetate or paper used in the lesson. Providing measuring units such as paper clips or small counting blocks will help build fine motor skills as children measure. Students can arrange the measuring units, end to end, along the trail and then count the distance traveled by the snail. They can measure several different trails to determine which snail traveled the farthest. If you're lucky enough to have snail trails on your school sidewalks or playground, encourage children to measure these trails, too.

Other Connections

Child's Life Connection

Students love the companionship and responsibility of having a pet, and there are few pets that require less expense than our friend the garden snail. Consider keeping a snail as a class pet or, with parent approval, giving each of your children a pet snail of their own. Your snail will need a ventilated container *with a lid* and plenty of leafy vegetables and plants to eat. Ensure that your snail's environment is moist and clean it regularly to remove snail poop and mucus. Encourage children to experiment with different foods to find what their snail likes best. Snails can be hand-fed; have a child gently hold a piece of food in front of the snail's mouth. If the snail is interested, it will begin to eat. You can feel (and hear) the snail's specialized hard, rough tongue grind the food as it eats. Uncooked oats are a snail favorite.

Center Connections

At the **art center**, provide children with the book *Swirl by Swirl: Spirals in Nature* by Joyce Sidman (2011) or *Snail, Where Are You?* by Tomi Ungerer (2015) and have children look for the different artistic interpretations of the snail shells, locating spirals in waves, fern leaves, and more. Students can then make different types of spiral artwork (e.g., a self-portrait using spirals for eyes, ears, nose, and mouth). At the **sensory table**, allow children to explore different slippery textures, such as petroleum jelly, vegetable oil, and dish soap. Prompt children to compare the different slimy substances and describe them as thick, runny, slippery, sticky, etc. Students can explore rubbing two blocks together with and without a slimy substance between them to determine which substance seems to make the blocks slide most smoothly. For **dramatic play**, use the book *The Snail's Spell* by Joanne Ryder (1988) to inspire children to imagine becoming a snail. This dramatic play will allow children to explain what they have learned about snails as they follow the book and act out being snails themselves. Ask children to decide what materials they need to serve as snail shells and tentacles or eyestalks and be ready to provide blankets or boxes (shells) and chopsticks or unsharpened pencils (tentacles or eyestalks) to help children pretend to be snails.

Family Activities

This week we've been studying a fascinating animal: the garden snail. You can ask your child about his or her snail observations and snail experiment. And you can further his or her science learning by going on a snail hunt together. Let what your child knows about snails guide you in your exploration. In your yard or at a local park, discover together where snails live. When you find a snail, let your child show you the different parts of a snail's body. (If the snail you find is sealed in its shell, setting it in a few drops of water will usually bring it out, but don't submerge the snail in water—they breathe air.) Notice the conditions where the snail is living and then search in similar places for new snails. Be sure not to collect snails; instead, return each one to where it was found. (If your snail hunt is unsuccessful, you can go to your local library and find picture books of snails. Before reading, flip through the book and let the pictures guide your child in telling you what he or she knows about snails.)

Actividades Familiares

Esta semana hemos estado estudiando un animal fascinante: el caracol. Puede preguntarle a su hijo sobre sus observaciones y experimentos en torno al caracol. Y puede contribuir aún más a su aprendizaje científico yendo de paseo en una búsqueda de caracoles juntos. Deje que su hijo y su conocimiento sobre el caracol lo guíen en su excursión. En su jardín o en un parque cercano, descubran juntos donde viven los caracoles. Cuando encuentre un caracol, permita que su hijo le muestre las distintas partes del cuerpo del caracol. (Si el caracol que encuentre está encerrado en su caparazón, por lo general salen al ponerlo en unas cuantas gotas de agua, pero no lo sumerja en agua—ya que respiran aire.) Tenga en cuenta las condiciones donde vive el caracol y luego busque caracoles nuevos en lugares parecidos. No atrape los caracoles; al contrario, devuelva cada uno donde lo encontró. (Si no tiene éxito en su búsqueda de caracoles, puede ir a su biblioteca local y buscar un libro de imágenes de caracoles. Antes de leer, avance y deje que las imágenes guíen a su hijo para que le diga lo que sabe sobre los caracoles.)

Assessment—What to Look For

- **Can children describe features of snails based on their observations?**
  (using senses to gather information, identifying properties)
- **Can children conduct and interpret the results of an experiment?**
  (carrying out investigations, constructing explanations, explaining based on evidence, using new or complex vocabulary)
- **Can children infer the purpose of behaviors such as snail mucus production?**
  (making inferences, identifying causality, using complex patterns of speech)

Standards

Head Start Early Learning Outcomes Framework
P-SCI 6. Child analyzes results, draws conclusions, and communicates results. "Analyzes and interprets data and summarizes results of investigation. Draws conclusions, constructs explanations, and verbalizes cause and effect relationships."
Next Generation Science Standards
Science and Engineering Practice: Planning and carrying out investigations. "Plan and conduct an investigation collaboratively to produce data to serve as the basis for evidence to answer a question. Make observations (firsthand or from media) and/or measurements to collect data that can be used to make comparisons."
Common Core State Standards for Mathematics
K.MD.A.2. Describe and compare measurable attributes. "Directly compare two objects with a measurable attribute in common, to see which object has 'more of'/'less of' the attribute, and describe the difference."
Common Core State Standards for English Language Arts
W.K.3. Text types and purposes. "Use a combination of drawing, dictating, and writing to narrate a single event or several loosely linked events, tell about the events in the order in which they occurred, and provide a reaction to what happened."

Swimming Fish

with Nicole Hawke

Lesson: Observing the external features and behavior of goldfish

Learning Objectives: Children will use vocabulary and drawings to represent detailed observations of goldfish and will use their observations as a basis for making inferences about the functions of different body parts.

Materials: Drawing and writing materials; magnifiers; goldfish, each in a large, clear container; fish food

Safety: Some bacteria associated with freshwater fish can be harmful to people. Ensure that children do not drink water from the fishbowl and that children thoroughly wash their hands with soap and water after each activity. Students with allergies or immune-system illnesses should not handle goldfish or the water in fishbowls. Goldfish are non-native species and should never be released into local streams, ponds, lakes, or wetlands. Make sure any spilled water is wiped up to prevent slipping hazards. Remind children to keep away from electrical outlets when working with water.

Teacher Content Background: Goldfish are omnivorous, freshwater fish related to common carp. Goldfish can grow to be more than 10 inches long. Although most people picture a generic orange fish, there are actually over 100 different varieties of goldfish that vary in several interesting ways. Goldfish can have three different types of scales (metallic, calico, and matte) and therefore come in a range of colors. There are also different fin varieties, including single caudal (tail) fin, twin caudal fins, and twin caudal fins with no dorsal fin. Even the eye type varies—goldfish can have normal, telescopic, upward-facing, or "water bubble" eyes. Goldfish are poikilothermic, meaning that their internal body temperature changes as the external temperature of their environment changes. The body temperature of goldfish is roughly the same as the water they are in. This characteristic means that you must take caution when changing their environment to prevent sudden changes in temperature. Despite common practice, goldfish actually do better in larger spaces with plenty of room for swimming. At a minimum, use a 20-gallon tank; use a much larger tank if you have more than one goldfish. With proper care, goldfish can live for 25 years or more.

Science terms that may be helpful for teachers to know during this lesson include *investigate*, *observe*, *fin*, *tail fin*, and *operculum*.

Procedure

Getting Started

Prior Knowledge: To begin this investigation, ask children to share what they know about goldfish and their personal experiences with goldfish. Then invite children to describe a goldfish to you. As they are discussing their mental image of a goldfish, record the different characteristics children mention (can swim, has fins and a tail, lives underwater, and so on) using words with picture support.

Initial Explanation: Ask the children to draw a picture of a goldfish from memory. Encourage them to include as many details as they can. As children draw, refer back to the list of goldfish characteristics and ask prompting questions such as "Where does a goldfish live? What color is it? What does its body look like? What is it doing?" Provide ample time for the children to draw their goldfish. As they draw, continue to ask individual children prompting questions to encourage more detail in their pictures. For example, if a child does not include fins or a tail, you could ask "How does your goldfish move?" To promote children's oral language development, ask children to describe their completed drawing aloud to a classmate.

[Note: Later in this lesson, children will be comparing their initial drawing to one they make based on observations of an actual goldfish. Title this first drawing "My Idea" and glue it to one half of a larger sheet of paper. On the other half of this paper, children will make a drawing titled "My Observation" that shows details of the goldfish they will observe later in the lesson. Having both their goldfish from memory and the goldfish they observed on the same paper will make children's comparisons of the two drawings easier.]

Investigating

Observing: Before giving children the opportunity to observe a real goldfish, briefly discuss safety and model how to make an appropriate observation of the goldfish. Communicate to children that they should not touch the water or the goldfish and that they should be careful to not move the containers to avoid spilling and possibly harming the goldfish. Bring out the container(s) of goldfish for the children to observe. Give children magnifiers and time to observe whichever aspects of the goldfish are of interest to them. After allowing children to observe freely, link back to the ideas children shared about goldfish earlier, such as ideas about body parts, habitat, and movements. Prompt children to see if they can identify any of the goldfish's body parts (fins, tail, eyes, mouth, operculum [or gill cover]). Ask the children if they see their goldfish moving. Ask them to describe to you what body parts seem to help the goldfish move.

Documenting: Once children have had an opportunity to observe the goldfish, ask them to make a drawing of their observation; emphasize to children that they should be drawing the actual goldfish in their container. Help children to pay special attention to their goldfish by asking guiding questions such as "What shape is your goldfish's body? What parts do you see on your goldfish's body? Does your goldfish have eyes? How many eyes? How many fins does the goldfish have? Where are the fins on the goldfish's body? How does your goldfish move? How do the fins move? How does the tail move?" Have children label their drawings "My Observation."

Making Sense

Describing Findings: Once children have completed their observational drawings, lead the class in a discussion about features common to all the goldfish (fins, tails, eyes, mouths, opercula, etc.). By highlighting the similarities among the goldfish in the class, children begin to realize what characteristics all goldfish have in common. Then have children place their initial drawings of goldfish from memory side by side with their observational drawings. Prompt children to look for similarities and differences between the two drawings. Discuss which picture looks more like the real goldfish. Let children explain why their second drawing is more accurate and reinforce the idea that in science, observations are important because they give us more accurate information and help us to notice important details.

Generating Explanations: After observing and describing their goldfish, children should feel comfortable using the words *fins* and *tail* and should be able to identify those parts on a goldfish. Some children may already be beginning to generate explanations about how and why fins and tails are used. Encourage this thinking through questions such as "All the goldfish had fins; what do you think they use their fins for? We noticed that all the goldfish had tails that moved back and forth, what do you think they use their tails for?" Encourage children to continue to observe goldfish structures and behaviors and to speculate about their purposes.

What's Next?

Extension Activity

After observing goldfish, ask the children, "How should we care for our goldfish?" Discuss the things the class will need to do in order to keep their goldfish healthy. If children are uncertain about how to keep a goldfish healthy, make connections between what the children need to be healthy and what goldfish need (see *Animal Walk,* p. 131, for information about the basic needs of animals). As a class, create a schedule that will be followed to care for the goldfish. The schedule should include feeding the goldfish small amounts of food twice a day. [Note: You will also need to include fish tank maintenance (such as changing the water, cleaning the filter, etc.), but that will depend on your individual setup.] Students should take turns sharing in the responsibilities of caring for the goldfish. Have children observe as the goldfish is being fed. Ask questions to help direct observations, such as "What does the goldfish do when we drop food in the water? How does the goldfish get to the food? What body parts help the goldfish get the food?" By encouraging the children to observe the goldfish eating, they gain firsthand evidence of the purposes of the fins, tail, mouth, and eyes. Once children have made and discussed observations of the goldfish feeding, encourage them to use the observations as a basis for other ideas. Ask children thought-provoking questions such as "Do you think all goldfish get their food the same way our goldfish does? Do goldfish in the wild get food this way?" This type of observation-based speculation is important for science and for generating even more interest and curiosity.

Integration to Other Content Areas

Reading Connections

Throughout the lesson, there are many ways to support children's literacy skills. Speaking and listening to each other helps children to begin to understand turn-taking during conversation, allows children to practice active listening, and develops language fluency. Further develop children's literacy skills and science learning by providing several different types of books related to fish. Through the use of informational books about fish and their body parts, such as *About Fish: A Guide for Students* by Cathryn Sill (2017), *Animals Called Fish* by Kristina Lundblad and Bobbie Kalman (2005), *Fabulous Fishes* by Susan Stockdale (2012), and *What's It Like to Be a Fish?* by Wendy Pfeffer (2015), children will have the opportunity to experience the features of informational text and to learn more about fish. Sharing quality literature about fish or fish characters, such as *The Pout-Pout Fish* by Deborah Diesen (2013), *Not Norman* by Kelly Bennett (2008), and *The Rainbow Fish* by Marcus Pfister (1999), can help children to see that information in science can also be used in creative ways when we use our imaginations.

Writing Connections

Encouraging children to record their observations supports their writing development; extend this by having children add labels to their goldfish drawings. Provide each child with labels for tail, fin, eye, mouth, and operculum. Support children as they identify each label and apply it to their drawing; ask questions such as "The word *tail* starts with which letter? So which of these labels is the word *tail*? Where on your drawing should you put the *tail* label?" Adding labels to their observations will help children to understand that print contains meaning and engages children in scientific communication using drawings and written words as tools to share what they have learned.

Math Connections

Creating a schedule of care for the goldfish (see the extension activity) links naturally to talking to children about time and measurement. With your children, create a daily, weekly, or monthly schedule. Set up two times per day that children will feed the goldfish and help children to learn how to tell these times on a clock. Establish a day of the week for changing the water in the tank, and reinforce days of the week and weeks in a month during calendar time. "We change the goldfish's water every Friday; look at the calendar and count how many times we'll change the water this month." Measurement can also be part of goldfish care. Students can count out or even weigh the number of food pellets the goldfish is fed. And when the goldfish's water is changed, children can learn different ways to measure amounts of water.

Other Connections

Child's Life Connections

From Elmo's Dorothy to Disney's Nemo, children are exposed to many fish characters. Ask your children to think of "famous" fish that they know from books, television, or movies. Have them identify which fish are real and which are pretend. Have conversations with children about the differences between real and fictitious fish and also about the characteristics that they have in common. Comparing pictures or watching short video clips of fictitious fish and of their real-life counterparts (e.g., Nemo and a real clownfish) can help facilitate this conversation.

Center Connections

At the **art center**, provide children with a variety of pictures of fish from old magazines. Provide paper, glue, scissors, and paint. Help children paint a background for their fish (this can be a natural setting full of other sea life or a simple fishbowl). Students can select a fish, cut it out, and then, once the background is dry, glue their fish picture in place. (For younger children, precut the pictures of fish.) At the **sensory table**, add rubber or plastic fish to the water. Allow children to act out behaviors they observed their fish doing. As children play with the fish, encourage them to describe what their fish is doing and the different parts of their fish, applying the vocabulary they learned during the lesson. For **dramatic play**, consider setting up a fish pet store. Include a variety of supplies one would need to care for a fish: containers, food, rocks, tank decorations, nets, and pretend fish. Include a cash register, price signs, and a vest or uniform for workers. Through this store, children can practice using the information they learned about fish and the ways to care for them.

Family Activities

At school, your child has been investigating fish. As a family, visit a place that has fish, such as a pet store or aquarium or a business with a large tank display. Talk to your child about the fish she or he sees. Ask your child to count the fish, to identify different body parts of the fish (fins, mouth, eyes, tail, etc.), to explain how the fish are moving, to compare two different fish, and to discuss anything else about fish that your child wonders about. Let your child's curiosity lead you to the library or internet to learn more about fish together.

If you are willing, try getting a fish as a pet. Have your child create a list of supplies the fish will need, and then allow your child to pick out the fish from the pet store. In your home, place your new pet in a place that is visible and accessible to your child. Create a schedule of care with your child, and help her or him to be responsible for the fish's care.

Actividades Familiares

En la escuela, su hijo ha estado investigando sobre el pez. Como familia, visiten un lugar que tenga peces, como una tienda de mascotas, un acuario o una empresa con un gran tanque en exhibición. Hable con su hijo sobre los peces que ve. Pídale a su hijo que cuente los peces y que identifique las distintas partes de su cuerpo (aletas, boca, ojos, cola, etc.), para explicar como se mueve el pez, para comparar dos peces que sean distintos y para discutir cualquier otra información sobre el pez que su hijo quiera saber. Deje que la curiosidad de su hijo lo guíe a la biblioteca o a la Internet para aprender más sobre el pez juntos.

Si tiene la disposición, intente adoptar un pez como mascota. Pídale a su hijo que elabore una lista de elementos que el pez va a necesitar y luego lleve a su hijo a elegir al pez a la tienda de mascotas. En casa, ponga a la nueva mascota en un lugar que sea visible para su hijo y al cual tenga acceso. Establezca un horario de cuidado con su hijo y ayúdelo a ser responsable del cuidado del pez.

Assessment—What to Look For

- **Can children describe fish based on their observations?**
  (using sense to gather information, identifying properties, using complex patterns of speech)
- **Can children compare their initial and observational drawings of a fish?**
  (comparing properties, explaining based on evidence, using new or complex vocabulary)
- **Can children infer the purposes of different body structures?**
  (making inferences, identifying causality, using complex patterns of speech)

Standards

Head Start Early Learning Outcomes Framework
P-SCI 1. Child observes and describes observable phenomena (objects, materials, organisms, events, etc.). "Makes increasingly complex observations of objects, materials, organisms, and events. Provides greater detail in descriptions. Represents observable phenomena in more complex ways, such as pictures that include more detail."
Next Generation Science Standards
Science and Engineering Practice: Analyzing and interpreting data. "Record information (observations, thoughts, and ideas). Use and share pictures, drawings, and/or writings of observations. Compare predictions (based on prior experiences) to what occurred (observable events)."
Common Core State Standards for Mathematics
1.MD.B.3. Tell and write time. "Tell and write time in hours and half-hours using analog and digital clocks."
Common Core State Standards for English Language Arts
W.K.3. Text types and purposes. "Use a combination of drawing, dictating, and writing to narrate a single event or several loosely linked events, tell about the events in the order in which they occurred, and provide a reaction to what happened."

Critter Camouflage

with Lauren M. Shea

Lesson: Exploring animals' use of camouflage to blend into their surroundings

Learning Objectives: Children will observe and compare how various animals can be camouflaged on different surfaces and in different environments. They will discuss why animals camouflage themselves and communicate their learning through speaking, writing, drawing, and acting.

Materials: Trays with various materials and toy animals (see possible environments in the table below); color photographs of real animals that camouflage (corresponding to toy examples in the table below); writing and drawing materials; large picture of a bird of prey (e.g., hawk, eagle, falcon, owl); a painted forest-like environment (mural) or use of a colorful wall; green or brown blanket

Safety: Be aware of and talk about potential choking hazards for children with some of the smaller plastic animals.

Teacher Content Background: Camouflage is an important tool for survival for many animals. Camouflage is an adaptation that allows an animal to blend into its environment. Many animals use camouflaging as a way to protect themselves from predators, whereas many predators use camouflage as a way to conceal themselves in an effort to ambush their prey. Animals camouflage themselves in various ways: perhaps the most common tactic is *background matching*. When an animal blends into the environment by resembling the colors, forms, or movement of its surrounding, it is camouflaging by background matching. For example, polar bears, snowy owls, and snowshoe

Table 2.1

Possible Environments

Color	Material	Toy Animals
Green	Leaves or grass	Frogs, grasshoppers, emerald butterfly
Brown	Sticks or woodchips	Turtles, toads, snakes
Beige	Sand or rocks (with or without water)	Sand crabs, fish (bottom dwellers), octopuses, or lizards, snakes, crickets
White	Snow (cotton balls)	Polar bear, snowshoe hare, Arctic fox, snowy owl

hares have fur that appears white to blend in with the snow that surrounds them. When animals use camouflaging to appear to be something else, it is called *disguising* (a walking stick insect looking like a stick or a twig) or *mimicry* (the nonpoisonous viceroy butterfly mimicking the look of the poisonous monarch butterfly). These various types of camouflage all aid in the survival of the animals that employ them. Amazingly, some animals, such as chameleons and octopuses, are capable of actively changing their skin patterns and colors to camouflage themselves as they move from one type of background to another.

Science terms that may be helpful for teachers to know during this lesson include *observe, camouflage, predator, prey,* and *environment*.

Procedure

[Note: Prior to the lesson, set up several trays with different materials (see *Materials*) and place different toy animals on each. For *Getting Started*, place animals that are camouflaged on each tray. For *Investigating*, use trays that have some camouflaged animals and some animals that stand out in the environment.]

Getting Started

Introduction: Show children a photograph of a bird of prey. Ask children to turn to a neighbor and tell them one thing they notice about that bird (the children might notice its color, its size, its talons, or its beak). Tell the children the name of the bird and ask them what they think this bird may eat (based on their observations and prior knowledge). Tell them that the photograph shows a bird of prey and tell them that birds of prey are birds that hunt and eat other animals. Tell children that many birds of prey have excellent eyesight and use this to see animals on the ground as they fly high above. Set out three trays, each with a different environment (e.g., green leaves and grass, sand and rocks, and sticks and soil, each with camouflaging animals hiding inside). Describe these environments and tell the children that the bird will now fly over each, using its excellent sense of sight to look for animals to eat. Ask the children to look closely at the surfaces to see if they can observe anything in the areas. As the children look closely, "fly" the bird over the environments. Let the children share what they notice about the surface and the animals inside.

Prompting Questions: When several children have visually discovered animals hiding in the trays, ask them what they know about this hiding. Ask prompting questions such as "What do you think the animals are doing? Why might they do this? How does this help the animal? What does this mean for the bird? What do you notice about all the animals in the same environment? Do you know of any other animals that do something like this? What would happen to an animal that was _____ [a contrasting color (e.g., a black bug on a sandy surface)]?" Record the children's answers in a visible place, using words with picture support.

Investigating

Observing: Encourage your children to become careful observers. Have them use only their sense of sight, like the bird of prey, to look for prey. Ask them to notice which animals are hiding well. Encourage children's thinking and curiosity by asking open-ended questions such as "What do you think the bird would see? What do you notice about the color of the surface and the color of the animal? What about this animal makes it stand out? How does this animal's color help or not help it in these leaves? How is this animal different from another animal on this surface? What are some things that might happen to this animal if it stays here? Why are some animals hard to see? What about the animals made them hide well?

Why might that quality be good for the animals? What are the ways these animals are blending into their surroundings?" Remember that these conversations serve to engage children's interest in the hiding abilities of animals; noting and celebrating children's ideas about the characteristics that help or hinder animals will support their excitement and learning about the topic.

Making Sense

Describing Findings: After all the children have had turns investigating the animals on the various surfaces, have them tell a friend about one animal and where it hid the best. Then prompt children to share ideas with the class by asking, "What did you find out? What was the best surface for the [green grasshopper] to hide? Would the [grasshopper] hide well in the sand environment? Why or why not?" Encourage the children to share their learning with a friend, encouraging them to be good speakers and listeners. After children discuss their observations (i.e., after developing an understanding of the science concept), you can then introduce science vocabulary. Tell the children the science word for when animals blend in with their surroundings is *camouflage*. Ask children if they have heard this word before and ask them to share their background knowledge of this word. Encourage children to repeat this new word; ask them to say the word *camouflage* in many different animals' voices. For example, they could say *camouflage* like a white polar bear in the white snow (deep voiced, "camouflage"), a green grasshopper in the green grass (tiny voiced, "camouflage"), or a brown owl in a brown tree trunk ("cam-whooooo-flage").

Application: To further solidify their learning, children need an opportunity to use the new concept and vocabulary they have learned. [Note: You will need a large picture of a forest environment with different shades of green and brown and other natural colors—use a poster or a piece of fabric with tree branches and leaves, or perhaps your class can paint a mural as a group art project.] Give each child a photograph of an animal that camouflages in the forest and, while looking at the mural, have the children decide where their animals would best camouflage themselves. Have the children individually place their animal photograph on the mural and, as they do so, explain to the class why they think that the animal would best hide in the location they chose. Throughout this process, encourage children's use of the new word *camouflage* and expand children's thinking by showing other places an animal might be placed, and asking them whether it would be a safe hiding place for it, and inviting children to make comparisons of two animals that hide best in the same area and of animals that hide best in different areas.

What's Next?

Extension Activity

To continue the conversation about camouflage, have a child (preferably choose a child who is wearing bright clothing) stand in front of the "forest" used in the Making Sense section of the lesson. Ask the children, "What do you notice about [your classmate] standing there? How well does she or he blend in? If [the child] wanted to hide in this forest, would standing there be a good hiding place? Why or why not?" Next, cover your volunteer with a brown or green blanket that matches the background and ask the children what they notice about him or her blending in now and how the blanket helped him or her to hide. Have all children try camouflaging themselves, then connect children's multiple experiences with camouflaging by asking questions such as "What have we noticed about the many ways animals can hide themselves?"

Have a day where the children come dressed up in a way to blend in with the class mural. Have the children stand in front of the painting and talk about how they blend in. Take a class picture and display it for others to find the children. Send a copy of the picture home so children can show their families and explain to them what is happening and why many animals do something similar.

Integration to Other Content Areas

Reading Connections

Informational books portraying photographs, pictures, and drawings of different animals that use camouflage, will help children to gain an interest in other creatures to study. Books such as *Animal Camouflage* by Vicky Franchino (2015), *Can You See Me?* by Ted Lewin (2015), *How to Hide a Butterfly and Other Insects* by Ruth Heller (1992), *Where in the Wild?* by David Schwartz and Yael Schy (2011), *I See Animals Hiding* by Jim Arnosky (2000), and *What Color Is Camouflage?* by Carolyn Otto (1996) can provide children with examples of animals blending in with their environments and disguising as other animals. Through these and similar texts, children will continue to understand how books can provide a source of knowledge. As you read aloud, ask questions about the hiding animals, prompting children to predict and then "read to find out." At these early ages, "reading to find out" is a fundamental and important idea for children.

Writing Connections

Go on a nature walk with the children to search for camouflaging animals both large and small. As you walk, inspect the grass, leaves, or sand on the ground; look under rocks and logs. Ask the children to report what they notice. Record the numbers of the different animals you find (see *Math Connection*). After the nature walk, create a class book so that others can read and learn about the experience. Each child can select one animal to write about. Help children to formulate their thoughts and, when their thought is concise, write the sentence using consistent sight words to promote their literacy development and concepts about print. For example, using a sentence such as "We saw a _____" for each animal will help the children start to connect initial sounds to letters and will support their recognition of beginning sight words. To go with their sentence, have each child draw a picture of the hiding animal in its environment. Then, glue the children's pictures and the words you wrote together onto large pages to create a big book for the class library. (This activity might spark interest in additional nature walks. See *Animal Walk*, p. 143, and other lessons in Chapter 4 for possibilities.)

Figure 2.4

Camouflage Data Sheet

Animal	Environment
Name:	Description:
Drawing:	Drawing:
How many did you see? ______________________	

A full-size version of this figure is available at *www.nsta.org/startlifesci.*

Math Connections

Data collecting provides opportunities for children to use their emergent math skills. While on the nature walk (see *Writing Connection*), children can carry a clipboard with paper to tally how many animals they find that are camouflaged in different environments. Create a data table and show children how neat data recording looks (see Figure 2.4). When you return to the classroom, model how the tallied data can be numerized and transcribed to a format that is easier for others to read. As a class, create a final poster of their findings from the nature walk.

Other Connections

Child's Life Connection

Children love to dress up. Tap into their desire to try on something new by asking them to put on clothes to blend in to a background or environment of their own choosing. Have them search out various colors around their homes, neighborhoods, and school, and then ask them to dress up in a way that would camouflage them against that background. If your school has a red (or other color) wall, have a "Red Wall Camouflage Day." Students can wear red clothes that day and pose in front of the wall for a camouflage photo. And maybe children have toys that are good at camouflaging. Students can search their stuffed animals and toys to find ones that camouflage against different backgrounds.

Center Connections

At the **art center**, children can draw animals of their choice and then create an environment in which it can camouflage. Provide children with real objects, such as sand, leaves, and sticks, to glue

onto their picture to allow children to match colors and feel the true textures where the animals hide. Provide children with photographs of animals that they have seen in the lesson so that children can model a particular animal's attributes in their drawings. At the **sensory table**, provide additional surfaces for the children to touch and explore to help support their further development of the idea of camouflaging. Students can play with a partner to hide and seek toy animals at the sensory tables. They can begin using their eyes (like the bird of prey) and then move on to their sense of touch to search throughout the table for the animals. In the **dramatic play** area, put out clothes of various colors. Students can dress to blend in with the different colors of the classroom (walls, carpet, etc.). Students may pretend to be predator or prey, using camouflage to hide from the other.

Family Activities

This week your child has been learning about one way that some animals protect themselves: camouflage. You can ask your child about his or her investigation of animals that blend into their surroundings. See what your child has discovered about the ways certain animals can hide themselves and the different reasons that an animal might want to hide itself. In your own neighborhood, you might be able to find some animals using camouflage to hide themselves. Go on a walk with your child and look through leaves, in soil, or on tree bark to observe animals. Ask your child what he or she notices about any animal that you find. He or she may tell you how color, shape, and size help the animal hide well. On another adventure, play Camouflage Hide and Seek with your child. Use camouflage to help you to hide and talk about how what you and your child are wearing might help you to blend in in one hiding spot and cause you to stand out in another. Ask your child to think about how your game of Hide and Seek compares to the animals that he or she learned about at school.

Actividades Familiares

Esta semana, su hijo ha estado aprendiendo sobre una forma en la que los animales se protegen: el camuflaje. Puede preguntarle a su hijo sobre su investigación acerca de animales que se mezclan con su ambiente. Vea lo que su hijo ha descubierto sobre las formas como ciertos animales se pueden esconder y las distintas razones por las cuales un animal podría querer esconderse. En su propio vecindario, podría encontrar algunos animales que utilizan el camuflaje para esconderse. Lleve a su hijo de paseo y busquen en las hojas, la tierra, o la corteza de árbol para observar animales. Pregúntele a su hijo qué es lo que observan de cada animal que encuentran. O le puede decir cómo el color, la forma y el tamaño ayudan al animal a esconderse bien. En otra aventura, juegue a las escondidas de camuflaje con su hijo. Utilice camuflaje para esconderse y hable sobre cómo lo que usted y su hijo vistan puede ayudarles a camuflarse en un lugar y causar el ser descubierto en otro. Pídale a su hijo que piense acerca de cómo su juego de escondidas se compara a los animales sobre los que ha aprendido en la escuela.

Assessment—What to Look For

- ***Can children describe the features of an animal that enable camouflaging?***
  (identifying properties, comparing properties, using new or complex vocabulary)
- ***Can children interpret the purpose of blending into an environment?***
  (identifying patterns, making inferences, constructing explanations, using complex patterns of speech)
- ***Can children communicate their learning through speech, writing, or drawing?***
  (using new or complex vocabulary, listening to and understanding speech)

Standards

Head Start Early Learning Outcomes Framework
P-SCI 2. Child engages in scientific talk. *"Uses scientific content words when investigating and describing observable phenomena, such as parts of a plant, animal, or object."*
Next Generation Science Standards
Science and Engineering Practice: Planning and carrying out investigations. *"Plan and conduct an investigation collaboratively to produce data to serve as the basis for evidence to answer a question. Make observations (firsthand or from media) and/or measurements to collect data that can be used to make comparisons."*
Common Core State Standards for Mathematics
1.MD.C.4. Represent and interpret data. *"Organize, represent, and interpret data with up to three categories; ask and answer questions about the total number of data points, how many in each category, and how many more or less are in one category than in another."*
Common Core State Standards for English Language Arts
L.1.5.C. Vocabulary acquisition and use. *"Identify real-life connections between words and their use (e.g., note places at home that are cozy)."*

Spiderwebs

with Kristin Straits

Lesson: Looking closely at spiderwebs

Learning Objectives: Children will observe spiderwebs to discover that spiderwebs come in different sizes, patterns, and shapes. Children will discuss the fact that some spiders use webs to capture prey, and children will simulate the process.

Materials: Spiderwebs (capture your own using spray paint and construction paper [see below], or print pictures of different spiderwebs), writing and drawing materials, magnifiers, a spray mister (filled with water and adjusted to produce a fine mist), two different colors of yarn, masking tape

Safety: Only about 1% of the 3,000 species of spiders found in North America have bites that are harmful to humans, but bites from these noxious species of spiders can have severe consequences and require immediate medical attention. Always observe webs from a safe distance and avoid contact with spiders; advise children to do likewise. Also, allergies to spiders and their webs are possible. Find out if any of your children have these allergies and familiarize yourself with the symptoms of allergic reactions in children. It is recommended that additional adults be available to help supervise children while outside observing webs.

Teacher Content Background: Although all spiders produce silk, not all spiders spin webs. Instead of spinning webs, spiders such as wolf spiders, jumping spiders, and tarantulas actively hunt, stalking and pouncing on their prey. However, many of the estimated 40,000 species of spiders do spin webs and the majority of these use their webs for catching food. Webs come in many different shapes, and scientists often classify spiders by the type of web they weave. Two of the most common types of webs are orb webs and tangle webs. Orb webs are the type that we most often think of. These flat, wheel-like webs, built by spiders such as garden spiders, are suspended in open areas to capture flying insects. Although some spiders sit in their webs to monitor the capture of prey, many employ a "signal thread"—a strand of silk that that runs from the center of the web to where the spider is waiting at or just off the edge of the web. Tangled webs are often called cobwebs and can appear messy and shapeless. This jumbled appearance occurs because, unlike the flat orb web, these webs are three-dimensional. Spiders such as house spiders and black widows often place their tangle webs in bushes or, more noticeable to us, in the corner of a ceiling. There are, of course, additional types of webs, including sheet webs, funnel webs, woolly webs, and others.

Science terms that may be helpful for teachers to know during this lesson include *observe, compare, spider, web, predator,* and *prey.*

Procedure

[Note: Prior to starting this lesson, you'll need to print pictures of different spiderwebs to share with children or, better yet, collect spiderwebs yourself. Follow this procedure, away from children, to collect and preserve webs:

1. Make sure the spider has left the web.
2. Spray the front and back of the web with spray paint.
3. Before the paint dries, place a piece of thick construction paper behind the web and carefully move the paper forward making the web stick to the paper. Hold the paper to the web and, strand by strand, break the web free around the paper.
4. Be sure to use contrasting paint and paper; black paint on white paper or white paint on black paper both show up well.
5. You may want to spray the web and construction paper with a clear acrylic paint after it has dried to preserve it.
6. Try to collect several different types of webs.]

Getting Started

Prior Knowledge: Have a discussion with children to help them to activate their prior knowledge about spiderwebs and to grow interested in learning more about webs. Begin by asking children how they get their food. After several children have responded, ask, "Do animals get their food that way? How do animals get their food?" Again, encourage several children to respond. This type of conversation helps to situate the topic of interest (how spiders get their food), thereby promoting children's understanding of how the topic relates to the big picture (animals have different ways of obtaining food). Then direct children's thinking toward spiders, showing a picture of a spider and asking, "What about spiders? How do spiders get their food?" When the conversation turns to spiderwebs, find out what children know about spiderwebs, asking questions such as "Where can you find spiderwebs? What do spiderwebs look like? What do spiderwebs feel like? Why do spiders build webs? How does a spider know when something is caught in its web?"

Investigating 1

Observing: Go on a walk to look for spiderwebs, specifically looking for webs without spiders. [Note: As with all field trips, explore the area beforehand to identify any learning opportunities to seize or hazards or distractions to avoid. Be sure to caution children to not touch or attempt to catch any spiders.] When you find a web, check to see if the spider is nearby, keeping in mind that some spiders hide beyond their web. If a spider is nearby, observe the spider and its movement for awhile before searching for another, unoccupied, web. Once you find a deserted web, gather children around to observe the web. Spray the web with water from a spray mister to make the web easier to see. Students will have excited conversations about what they see. This is the type of enthusiasm you're striving to generate in all your science lessons. When appropriate, expand children's conversations by asking questions such as, "What is the shape of the spiderweb? What is the web connected to? Is a spider in the web? Where do you think the spider goes when it is not in the web? How do you think the spider made its web? What makes you think so?" Give the children a magnifier to encourage closer inspection of a strand of the web. Invite children to touch

Touching a spiderweb

the web and encourage them to describe how it feels.

Investigating 2

Observing and Comparing: Back in the classroom, capture children's interest by asking, "Do you know that I have a spiderweb collection? Would you like to see it?" Share sample webs that you have collected or printed from the internet for children to compare. Model for children how to compare webs and then ask children to work with a partner to compare two different webs. Prompt children to look for similarities and differences between the two webs, specifically encouraging children to consider the size, shape, and pattern of each web. Also, ask children what shapes they notice in the webs (see *Math Connections*). After children have compared the two spiderwebs, give them a third to consider. Ask them to decide which of the first two webs the third web is most like and to identify features that the third web has in common with the first two.

Making Sense

Generating Explanations: After children have described what they can observe about the webs, prompt them to take the next step and to infer different attributes of the different webs. Help children to analyze the webs from your collection; ask, "What kind of prey do you think a spider could catch in this type of web? Which web do you think catches the largest prey? Why? Which do you think would catch the smallest prey? Why?" Throughout this conversation, don't worry about right or wrong responses from children, but do encourage children to explain their thinking.

Application: Work together with your children to build a spiderweb out of yarn. Arrange your children in a circle and pass a ball of yarn across the circle from child to child. When you are finished, attach a strand of yarn (in a different color) to an outside edge of the web and weave it through the

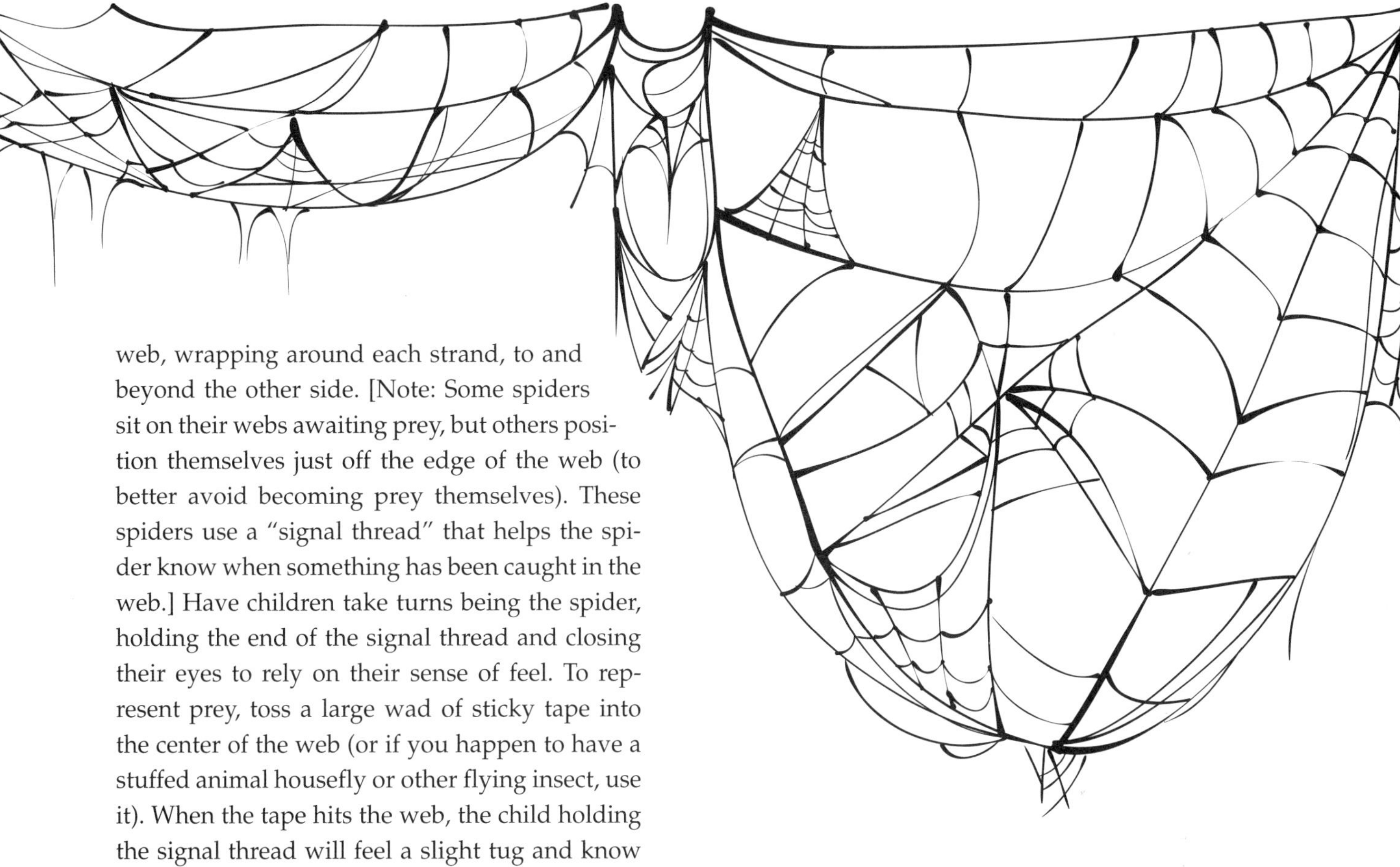

web, wrapping around each strand, to and beyond the other side. [Note: Some spiders sit on their webs awaiting prey, but others position themselves just off the edge of the web (to better avoid becoming prey themselves). These spiders use a "signal thread" that helps the spider know when something has been caught in the web.] Have children take turns being the spider, holding the end of the signal thread and closing their eyes to rely on their sense of feel. To represent prey, toss a large wad of sticky tape into the center of the web (or if you happen to have a stuffed animal housefly or other flying insect, use it). When the tape hits the web, the child holding the signal thread will feel a slight tug and know that something hit the web. If the child were a spider, it would run into the web to attack its prey. After the activity, ask children, "How do spiders capture their food? How do spiders know something has landed in their web?"

What's Next?

Extension Activity

Provide a picture of an orb web or show a time-lapse video of a garden spider constructing its web. Point out that the initial threads of webbing help to support the web and that after these supports are in place the spider spins a spiral of sticky webbing used to capture prey. Give pairs of children a small ball of yarn and allow them to make a spiderweb between the legs of a small chair turned upside down. For children looking to make a web that resembles an orb web, tie pieces of string diagonally from chair leg to chair leg to serve as support threads. Tie the yarn to the center of this *X* and allow children to weave their web by spiraling the yarn around each piece of string. [Note: Younger children will need additional assistance with this fine motor task.] Provide children with a toy spider and toy insects so children can act out how spiders capture prey using their webs. Finally, prompt children to draw a picture of, or take and print a photograph of, the web they created, adding labels as appropriate.

Integration to Other Content Areas

Reading Connections

Books are a great way to learn more about spiders and to see them up close. *Spiders* by Gail Gibbons (1993), *Spiders Have Fangs* by Claire Llewellyn (1997), *Spinning Spiders* by Melvin Berger (2003), *Are You a Spider?* by Judy Allen and Tudor Humphries (2003), *Next Time You See a Spiderweb* by Emily Morgan (2015), and many other books can give children additional insight into the world of spiders. Also, many books, such as *Spiders* by Nic Bishop (2012) and *Spiders* by Laura Murray (2016), include amazing, up-close photographs of these fascinating animals. To help children further understand what they learned during the science lesson, share with them books that show pictures of spiders making a web, such as *Spiders and Their Webs* by Linda Tagliaferro (2004).

Writing Connections

With your assistance, children can use a spider diagram (see Figure 2.5) to write eight words (one on each leg) that describe or relate to spiders. Eight words describing spiders may be too many for younger children, so consider completing the diagram as a whole class or in small groups. Also, children, using scribbling or inventive spelling, can write four words on the left of the spider and you can rewrite the words on the opposite leg. Or you can write four words on the left and children can rewrite your words on the right side of the spider. Younger children can write the numbers one through eight, one on each leg.

Math Connections

Draw spiderwebs on paper plates and put plastic or paper flies on each plate. Students can count the number of flies caught in each web and create addition problems to solve (e.g., "This web caught three flies. This web caught two flies. How many flies were caught in total?"). Spiderwebs are also great for exploring shapes. Give children crayons and a spiderweb diagram or drawing and ask them to find different shapes. Prompt them to color a triangle green, trace a spiral in red, color a circle blue, trace a zigzag line in yellow, etc. And spiders, with their eight eyes and eight legs, are great inspiration for counting and learning numbers one through eight.

Figure 2.5

Spider Writing

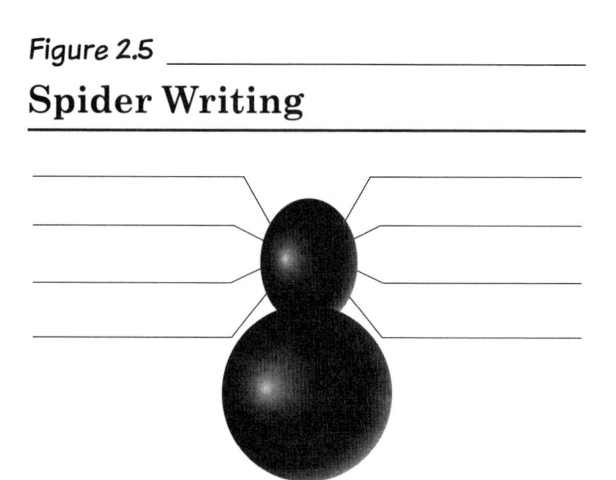

A full-size version of this figure is available at *www.nsta.org/startlifesci.*

Other Connections

Child's Life Connection

With adult supervision, children can go on a spiderweb hunt and look for the different sizes and shapes that make up webs. When a child finds a web, encourage him or her tell you about the web and what he or she notices. Ask the child questions such as "What shapes do you see in this web? What is the web connected to? Is there anything caught in the spider's web? Where do you think the spider is? What do you think the spider doing?" Have children draw the webs they see and invite them to share their drawings, explaining the web and what they know about spiders, to their classmates.

Center Connections

At the **art center**, children can build spiders out of Styrofoam and pipe cleaners. Spiders have two body parts—the head (technically called the cephalothorax) and the abdomen—and eight legs connected to the cephalothorax, but allow for more "creative" designs. Let children make the spider's body (two different sizes of Styrofoam) and attach the spider's legs (eight pipe cleaners attached to the smaller piece of Styrofoam). Your children can give their spiders more detail by adding eyes (spiders typically have eight, two larger eyes and six smaller) and a mouth with fangs. Make one or two spiders yourself to serve as models for your children. At the **sensory table**, put out different "threads" for children to explore. Use threads of different thicknesses, textures, and strengths (e.g., yarn, twine, string, rope, etc.). Students can discuss which material they think would make the best spiderwebs and why. In **dramatic play**, children can pretend to be spiders or flies negotiating a web. Use masking tape to create a large web on the floor. Students can walk across the web, either trying to stay on the strands of web (tape) or trying to only step between the strands of web. Set out large toy spiders and insects or clothes for dress-up (e.g., a spider hat [a hat with four "legs" hanging from each side], butterfly wings, etc.).

Family Activities

Your child has been investigating spider webs at school; you can continue this investigation at home. With your child, find spider webs at your home or pictured in a book. Use the webs you see to inspire the two of you to draw spider webs together. As you draw ask your child for advice, letting her or him tell you what she or he knows about webs and how they're made. You can also draw, paint, or create spiders together. Search the internet for spider craft ideas. Present a few options to your child and let her or him choose which spider she or he wants to create. As you make your spiders, remember that they have eight eyes and eight legs. Celebrate the number eight—serve your child eight apple slices or eight carrot sticks as a snack, or look with your child for other examples of things that come in sets of eight and count them together (for example, "Our home has eight doors! Let's find eight red blocks!").

Actividades Familiares

Su hijo ha estado investigando sobre las telarañas en la escuela; usted puede continuar con esta investigación en casa. Con su hijo, busque telarañas en casa o en un libro. Utilicen las telarañas que encuentren para dibujar telarañas juntos. A medida que usted dibuje, pídale consejos a su hijo y deje que le muestre lo que sabe sobre las telarañas y cómo se forman. También pueden dibujar, pintar o crear arañas juntos. Busque en la Internet por ideas de manualidades con arañas. Preséntele a su hijo algunas opciones y deje que elija qué arañas quiere hacer. A medida que haga las arañas, recuerde que tienen ocho ojos y ocho patas. Celebre el número ocho—sírvale a su hijo ocho rodajas rebanadas de manzana y ocho palitos de zanahoria, busque con su hijo otros ejemplos de grupos de ocho y cuéntenlos juntos (por ejemplo: "¡Nuestra casa tiene ocho puertas! ¡Busquemos ocho bloques rojos!).

Assessment—What to Look For

- **Can children describe spiderwebs based on their observations?**
  (using senses to gather information, identifying properties, using complex patterns of speech)
- **Can children compare different spiderwebs?**
  (comparing properties, explaining based on evidence, identifying shapes and patterns)
- **Can children infer possible prey based on attributes of spiderwebs?**
  (making inferences, identifying causality, using complex patterns of speech)

Standards

Head Start Early Learning Outcomes Framework
P-SCI 1. Child observes and describes observable phenomena (objects, materials, organisms, events, etc.). "Makes increasingly complex observations of objects, materials, organisms, and events. Provides greater detail in descriptions. Represents observable phenomena in more complex ways, such as pictures that include more detail."
Next Generation Science Standards
Science and Engineering Practice: Planning and carrying out investigations. "Plan and conduct an investigation collaboratively to produce data to serve as the basis for evidence to answer a question. Make observations (firsthand or from media) and/or measurements to collect data that can be used to make comparisons."
Common Core State Standards for Mathematics
K.MD.A.2. Describe and compare measurable attributes. "Directly compare two objects with a measurable attribute in common, to see which object has 'more of'/'less of' the attribute, and describe the difference."
Common Core State Standards for English Language Arts
L.1.5.B. Vocabulary acquisition and use. "Define words by category and by one or more key attributes (e.g., a duck is a bird that swims; a tiger is a large cat with stripes)."

Feeding Birds

with Kristin Straits

Lesson: Using feeders to attract and observe birds

Learning Objectives: Children will experiment with different foods to determine bird preferences. Students will record and analyze data, and use the data to inform choices as they create their own bird feeders.

Materials: Writing and drawing materials; photographs or other visuals of different local birds; clean empty milk cartons; heavy string; pencils or sticks; two jar lids; a tray or small board; an assortment of bird food such as birdseed, popped popcorn, leftover fruit, etc. [Note: Do not use uncooked rice as a possible bird food; ingesting it can be extremely harmful to birds.]

Safety: Be aware of different bird foods as potential choking hazards and of possible child allergies to several different materials, including feathers and nuts. Be sure to find out if any of your children have allergies before using the materials in this lesson and familiarize yourself with the symptoms of allergic reactions in children. Remind children not to eat any of the food used for the birds. Be sure to instruct children to not put their fingers in their mouth or nose while handling the bird food and make sure children thoroughly wash their hands with soap and water after the activity. Only use bird food free of pesticides and herbicides. Use caution when handling sticks as, they can be an eye hazard.

Teacher Content Background: Across the class of animals called birds, there is a great deal of variety; consider the ostrich, vulture, penguin, and hummingbird. One of the important ways in which birds vary is in their preferred food and the ways that they obtain this food. There are birds that eat seeds, worms, grass, fish, clams, mosquitoes, frogs, nectar, lizards, fruits, carrion, shrimp, rodents, roots, crabs, snakes, rabbits, and even other birds. A bird's bill gives us a hint about its food preferences. Birds such as finches, cardinals, and sparrows have relatively wide, cone-shaped bills—perfect for cracking open the seeds that make up much of their diet. Hummingbirds have long, slender bills for reaching deep into flowers to obtain nectar. The hooked bills of hawks, eagles, owls, and vultures make them ideally suited for tearing flesh. Warblers have small, pointy bills for picking insects off of plants. And with approximately 10,000 different species of birds in the world, the list goes on and on. (See *Looking for Birds*, p. 137, for more information on birds.)

Science terms that may be helpful for teachers to know during this lesson include *experiment*, *data*, *tally*, *bill*, and *seed*.

Procedure

[Note: *Looking for Birds*, p. 137, can serve as an introduction to this lesson.]

Getting Started

Introduction: Ask children what their favorite food is. Use pictures of different foods to encourage a variety of answers. Encourage responses from all children. "Does anyone have a different favorite food?" Keep track of all the different favorite foods and wrap up this discussion by pointing out that different people like different foods.

Prompting Questions: After discussing children's favorite foods, ask, "I wonder: do different kinds of birds like different kinds of food?" [Note: If you've already completed the *Looking for Birds* lesson, p. 137, you may want to revisit each of the specific birds your children have already seen.] "Do you think the blue jay likes the same kind of food as the house sparrow?" "What does the robin like to eat?" Ask children to support their responses by explaining their thinking. "Why do you think the blue jay and the house sparrow like the same foods?" "What makes you think the robin likes to eat worms?"

Investigating

[Note: Prior to *Investigating* you'll have to set up a platform feeder for your food choice experiment. Simply attach two jar lids of the same size to a small tray or board so that you have two small dishes a few inches apart from each other. Find a place outside where you can place your platform feeder, preferably a place that is near trees and shrubbery and is visible from the classroom window. Depending on how "birdy" your neighborhood is, you may want to put food out for birds for several days before you begin this lesson—birds may need time to find your feeder. Also, be mindful of squirrels as possible saboteurs of this lesson. If needed, use string to suspend your platform from a tree or structure (building, playground equipment, etc.). Let your children try out different techniques and locations to outsmart the squirrels.]

Experimenting: After asking children what they think the birds will like to eat, ask them how we might test these ideas. Show them your platform feeder. And ask children how we might use this and together create a plan for a "food choice" experiment. After deciding how to conduct your experiment, it's time to begin. Fill each lid with a different food. You can test just two alternatives or try several different foods over the course of several weeks; in addition to different kinds of commercial birdseed (sunflower, millet, nyjer seed, etc.), try popped popcorn, raisins, unsalted peanuts, small pieces of different kinds of fruits and vegetables, cereal, cooked rice or pasta (small pieces), and mealworms. "Which food do you think birds will prefer? Why?"

Observing and Documenting: If possible, set or hang your feeder so it is visible from the classroom windows. Check the feeder frequently to learn about the foods your visitors prefer. Which foods seem most popular? Does a particular type of bird seem to prefer a certain variety of food? You can set a schedule for feeding and observing birds or you can let children observe on their own terms during free time. Either way, you might want to create observation forms that have pictures of your common birds. Next to each picture have two boxes: one for each type of food for children to tally the number of visits (see Figure 2.6, p. 69). Be sure to include pencils in the area so children can record their data.

Making Sense

[Note: Prior to this activity, you'll have to collect empty milk cartons and cut away a section of the sides of each carton, making sure that the bottom inch of the carton is intact, as shown in the diagram. To make a perch, place a pencil or stick through small holes you have cut into the sides of the carton. Carefully punch a hole in the top of the carton, tie string through the hole, and make a loop for hanging.]

Application: Show children one of the milk carton bird feeders and describe how birds can fly up to the feeder, land on the stick, and poke their heads in to eat the food in the bottom of the feeder. Give each child his or her own feeder and allow them to decorate their feeder (see *Other Connections*). Review the data you've collected during the investigation and ask your children to identify the bird they like best. "Which bird do you want to visit your feeder? Based on our data, which food should you choose to attract that bird?" Let children fill their feeders and, if possible, hang them so they are visible from the classroom windows or send the feeders home so children can attract birds to their homes and share this activity with their families. When you send the feeders home, put the food in a sealable bag to help ensure that it makes it home, and remind the children that it may take several days for birds to find their feeders. They and their families should keep looking every day!

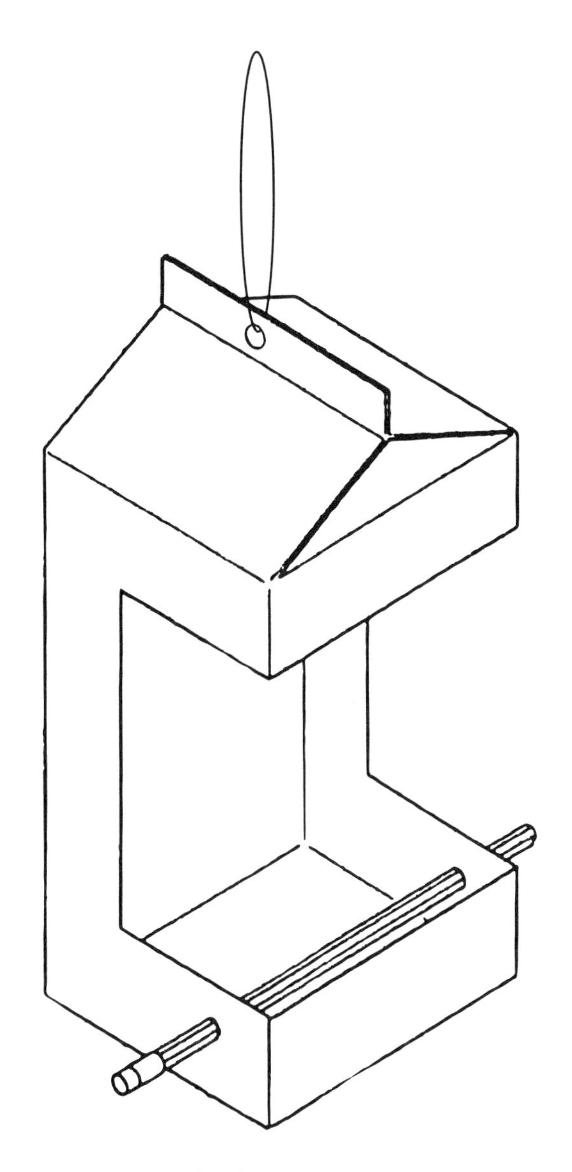

Milk carton bird feeder

What's Next?

Extension Activity

A variation of the experiment described above is to set out two different types of materials that birds could use in their nests. There are lots of different materials to consider, such as 4- to 8-inch lengths of yarn, ribbon, or string; hair, fur, or wool; feathers; cotton batting; strips of cloth; and shredded paper. Place one type of nest material on one side of the platform and a different material on the other side. Students can observe over a period of time to determine if different birds prefer different materials. [Note: Use this variation in the spring when birds are actively building nests.]

Bird on milk carton bird feeder

Integration to Other Content Areas

Reading Connections

Within the lesson and extension activity, there are several opportunities for children to acquire and use descriptive oral vocabulary; this vocabulary development supports emerging reading skills. During reading time, inspire your future and novice readers by sharing books related to your science investigation. Students can learn more about why different birds prefer different foods through books such as *Beaks!* by Sneed Collard (2002), *Unbeatable Beaks* by Stephen Swinburne (1999), and *Birds Use Their Beaks* by Elaine Pascoe (2001). You can also share books about birds' nests such as *Even an Ostrich Needs a Nest* by Irene Kelly (2009), *You Nest Here With Me* by Jane Yolen and Heidi Stemple (2015), *Mama Built a Little Nest* by Jennifer Ward (2014), and the beautiful photographs of real nests in *Nests* by Sharon Beals (2011). Additionally, books such as *Urban Roosts: Where Birds Nest in the City* by Barbara Bash (1992) and *The Tale of Pale Male* by Jeanette Winter (2007) can help children in urban areas become more interested in the birds they share their city with.

Writing Connections

Students can write a "lab report" using pictures and, if appropriate, words to describe the experiment from the *Investigating* section of the lesson. Title children's reports "Feeding Birds Experiment" and allow them to draw descriptions of how the experiment was set up and the results that were found. Provide children with a template, such as the one shown in Figure 2.7. Students can glue a small amount of each food they tested into the circles on their page and then draw the birds that visited each food. At the bottom of their reports children can write (or dictate to you) a sentence that summarizes the results of their experiment (e.g., "More birds ate the popcorn."). This type of writing helps children to understand that writing helps to communicate information and that writing is an important aspect of science.

Math Connections

Simple bar graphing offers another wonderful way to "tell the stories" learned through these feeding observations. For example, display a chart that shows the pair of foods or nest materials used in your experiment, and encourage the children to

Figure 2.6

Bird Observation Form

[name of common bird #1]	[food type #1]
[Insert picture of common bird #1 here]	
	[food type #2]
[name of common bird #2]	[food type #1]
[Insert picture of common bird #2 here]	
	[food type #2]

Full-size versions of the figures on this page are available at *www.nsta.org/startlifesci*.

observe carefully and report which food or material the bird visitor selects. Each report is recorded on the chart by adding a sample of that food (or a piece of the nesting material) to the appropriate "bar." For example, a piece of popcorn or sunflower seed is glued to the chart each time a child reports that a bird visitor made that choice. If both foods are selected, children can mark both "bars" and discuss this occurrence. When observations have been completed, ask the children to describe what they have learned about preferences at your bird feeder. Use questions such as, "Which food did the birds like the most? Which one did they like the least?" to encourage this conversation and be sure to help children use concepts of equality (e.g., more than, least, equal, etc.) as they describe their findings.

Figure 2.7

Feeding Birds Experiment

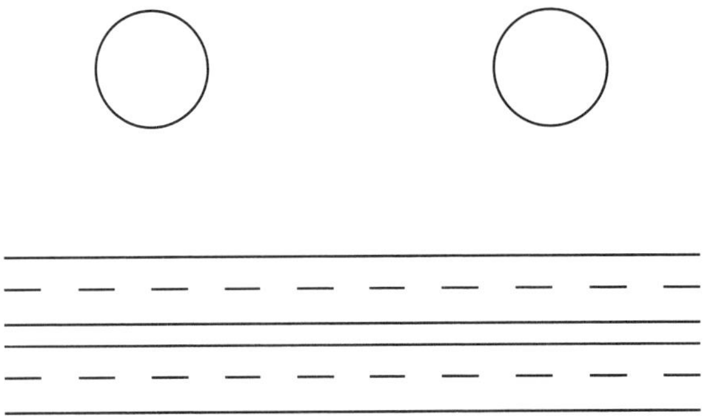

Other Connections

Child's Life Connection

Birds aren't the only ones with different food preferences. Students can test the preferences of a family pet or a family member. Brainstorm with children how they might conduct this food preference experiment at home. Which person or pet will be the subject of their experiment? Which two foods will be tested? What outcomes do the children anticipate? And children can be their own test subjects, trying two new foods to see which they like best.

Center Connections

At the **art center**, children can add personal touches to their bird feeders. They can paint the feeders as they wish and use natural materials to further decorate feeders—gluing leaves, feathers, sticks, and seeds to the feeder. Encourage children to decorate their feeders in a way they think will be most appealing to birds and ask children to explain how they decorated their feeder and why. Put out a "touch box" at the **sensory table**. (This is a box that children can reach, but not see, into.) In the box, put a bowl filled with different kinds of bird food (e.g., different types of birdseed, raisins, pasta, popcorn, etc.). Let children feel the bird food. Have them pretend that they are a bird that only likes one type of food and challenge them to collect that food using only their sense of touch. During **dramatic play**, provide bird costumes or large pictures of different birds that children can wear as they try picking up "food" with different "beaks" and carrying it back to their "nests." Put out a large bowl of items representing different foods (e.g., cotton balls, lima beans, uncooked macaroni, etc.). Give each child a smaller bowl to serve as their nest and provide several different tools for children to use as "beaks" to pick up the food. Possible tools include forks, chopsticks, spatulas, tongs, etc.

Family Activities

We've been feeding birds at school, and you and your child can feed birds at home. Repurpose household materials to make bird feeders. Once emptied and cleaned, many "trash" items can be converted into all sorts of bird feeders. Use the feeder that your child brought home from school as a model. Try different plastic bottles (water, detergent, milk, etc.), boxes, tubs, and other containers. Draw on your child's expertise; ask him or her to share the results of their investigation at school and to tell you which different foods you should use in your feeders to attract different birds. Make two different styles of feeders and study the birds that visit—do birds prefer one type of feeder to another? Ask your child how the two of you can make your feeder even more attractive to birds.

Actividades Familiares

Hemos estado alimentando pájaros en la escuela, y usted y su hijo pueden alimentar pájaros en casa. Reutilice materiales del hogar para hacer comederos para pájaros. Una vez vaciados y limpios, muchos artículos considerados como "basura" se pueden convertir en todo tipo de comederos. Utilice el que su hijo llevó a casa de la escuela como modelo. Pruebe con distintas botellas plásticas (agua, detergente, leche, etc.), cajas, tubos y otros recipientes. Aproveche la experiencia de su hijo; pídale que comparta con usted los resultados de su investigación en la escuela y que le diga los distintos tipos de comida que se deben utilizar en los comederos para atraer a distintos pájaros. Haga dos tipos de estilos de comederos y estudie los pájaros que visitan – ¿Los pájaros prefieren un tipo de comedero por sobreque otro? Pregúntele a su hijo cómo ambos pueden hacer que su comedero sea aún más atractivo para las aves.

Assessment—What to Look For

- **Can children ask questions about birds' food preferences?**
  (asking questions, identifying properties, making inferences)
- **Can children design an experiment to compare two variables?**
  (comparing properties, planning and carrying out investigations)
- **Can children record data about food preferences?**
  (documenting and representing findings, using computational thinking)
- **Can children use data to inform their decisions in designing their bird feeder?**
  (identifying patterns and causality, explaining based on evidence, using complex patterns of speech)

Standards

Head Start Early Learning Outcomes Framework
P-SCI 5. Child plans and conducts investigations and experiments. "With increasing independence, engages in some parts of conducting complex investigations or experiments. Uses more complex ways to gather and record data, such as with adult support, makes a graph that shows children's favorite snacks."
Next Generation Science Standards
Science and Engineering Practice: Planning and carrying out investigations. "Plan and conduct an investigation collaboratively to produce data to serve as the basis for evidence to answer a question. Make observations (firsthand or from media) and/or measurements to collect data that can be used to make comparisons."
Common Core State Standards for Mathematics
1.MD.C.4. Represent and interpret data. "Organize, represent, and interpret data with up to three categories; ask and answer questions about the total number of data points, how many in each category, and how many more or less are in one category than in another."
Common Core State Standards for English Language Arts
W.K.3. Text types and purposes. "Use a combination of drawing, dictating, and writing to narrate a single event or several loosely linked events, tell about the events in the order in which they occurred, and provide a reaction to what happened."

3

Plants

From towering eucalyptus to the smallest of grasses, flowering plants[1] represent a wide range of diversity. This diversity has captivated scientists and naturalists throughout the ages. What is just as fascinating is that despite this diversity, much about flowering plants is consistent—most have roots, stems, leaves, flowers, and (seed-bearing) fruits. The different forms these structures take help to account for the amazing diversity among plants. Engaging your children in meaningful explorations of plants can help them to experience some of the awe and excitement shared by Darwin, Muir, Linnaeus, McClintock, and countless others.

The lessons in this chapter will give you and your children opportunities to explore the diversity and the structures of flowering plants. Across these lessons, young learners will experience plants through multiple senses as they feel the texture of seeds and pumpkin guts, smell the scents of different leaves and other plant parts, taste new fruits and vegetables for the first time, and see a variety of different plants and plant parts.

A word of caution: As plants cannot run and hide from hungry animals, many have adapted chemical defenses to deter would-be herbivores. To humans, some of these chemicals—in small amounts—actually taste good and are used as spices in cooking. However, different people have different reactions to plant chemicals, and many allergic reactions to these chemicals are possible. Check with parents and guardians, as well as your school nurse, to find out if any of your children have plant or food allergies before using them in your classroom, and take time to remind yourself of the symptoms of allergic reactions in children.

1 Flowering plants represent just one subset of all plants. The other groups are cone-bearing plants (e.g., pines and cedars) and spore-bearing plants, which include nonvascular (e.g., mosses) and lower-vascular (e.g., ferns) plants.

Sorting Seeds

with Kristin Straits

Lesson: Observing, comparing, and sorting seeds or beans

Learning Objectives: Children will observe similarities and differences among a group of seeds or beans. Students will sort the seeds or beans based on their observations.

Materials: Writing and drawing materials; empty containers; magnifiers; several types, sizes, and colors of seeds or beans [Note: A bag of 18-bean soup mix may be purchased at most grocery stores and is useful for this activity. Also, throughout the lesson we describe the sorting of "seeds or beans"; use whichever term is appropriate given the materials you supply.]

Safety: Be aware of seeds as potential choking hazards and of possible allergies to different seeds or beans—avoid using peanuts and tree nuts. Be sure to find out if any of your children have seed or bean allergies before using these materials in your classroom and remind yourself of the symptoms of allergic reactions in children.

Teacher Content Background: Sorting is the act of grouping similar objects, ideas, or phenomena. Scientists sort the objects and events they study by arranging them into categories, bringing order to what might otherwise be an overwhelming amount of information. And in our everyday lives, it is often helpful and time saving to arrange objects into convenient categories. Because sorting things into categories plays such an important role in our lives, early experiences with sorting objects will help young learners to make better sense of their everyday world. It is important to understand that typically, a young child's ability is limited to being able to sort by just one characteristic at a time; moving beyond this comes only with practice and the child's further development of reasoning abilities.

Science terms that may be helpful for teachers to know during this lesson include *sorting*, *characteristic*, *seed*, and *set*.

Procedure

Getting Started

Introduction: Distribute an assortment of similar yet different types of seeds or beans to each child. Ask children to look closely at the seeds or beans and to share what they notice. Magnifiers will help children to make more detailed observations, especially of the smaller seeds or beans. Listen to what children say and observe what they do before asking questions.

Sorting shells

containers for each child's collection. Explain that grouping similar objects together is called sorting (i.e., sorting is the act of putting things into groups based on their characteristics). Be aware that *sorting* will be a new term for many of your children. As with all new vocabulary, you can help children to learn this term by repeatedly modeling its use and by encouraging your children to use it throughout the lesson. Ask the children to sort their seeds or beans according a specific characteristic that you choose, such as color, size, shape, texture, or smell. Give them time to carry out several practice sorts to allow them to become confident about the process. Then encourage children to sort based on properties of their choosing. As they sort, ask children to describe the similarities and differences among the seeds or beans. If individual children have trouble starting, describe what other children are doing—for example, "Julie is putting all the red beans together." Have each child share something about his or her seeds or beans and have children explain the thinking behind their sorting by asking, "How did you sort the seeds or beans? Why?" If time allows, ask children to work with a classmate and find a new way to group their seeds or beans together.

Prompting Questions: After children have had a chance to explore, encourage further observation with questions such as "Which seeds or beans seem to look alike? Which seeds or beans feel alike? How would you describe that seed or bean? What colors are the seeds or beans? Do any have dots on them? Are any wrinkled?" Ask the children to hold up their largest seed, smallest seed, most wrinkled seed, and so on.

Investigating

Comparing and Sorting: Ask children to group together all the seeds or beans that seem to be the same. Consider using small plates or trays as

Making Sense

Application: Although children observe and learn a lot about the physical characteristics of seeds or beans during this lesson, this lesson is primarily about sorting. After children have had opportunities to sort seeds or beans, they need a chance to use this new skill in a new context. Ask children to recall their sorting of seeds or beans and the different characteristics of seeds or beans. Tell them that next they'll sort something different. Give children a small collection of items (4–12, depending on the developmental age of each child) to sort. A collection of buttons or seashells will work perfectly. (We'll use buttons as our example below.)

To begin, supply each child with an assortment of buttons of varying designs, colors, and sizes. Encourage the children to examine the buttons with care and to describe some of the characteristics of each. After allowing enough time for preliminary exploration, invite children to sort the buttons in any way they might like. As they do so, ask them to describe why they have chosen to sort the buttons as they have. There is no one correct way to sort the buttons; point out to children that they can sort them any way they wish, as long as they can explain to others how their sorting was done. A very important aspect of successful sorting is being able to have other individuals understand how your sorting system works. Older children can analyze each other's sets of buttons to determine how they are sorted. This will help children to begin to understand that one of the important purposes of sorting is for better communication of information to others. [Note: Many other lessons also engage children in sorting. See, for example, *Nature Bracelets*, p. 144.]

What's Next?

Extension Activity

Provide each child with a small "fun-size" bag of M&Ms candy. Before they open their bags, raise a few questions about the contents and ask the children to make some guesses about the contents of their bag. "How many M&Ms do you think you will find in your bag? How did you come up with that number? Which colors do you expect to find in your bag? Which colors will you have the most of?" Encourage the children to display their responses on paper, with your help as needed. Have them carefully empty the contents of onto the sheet of paper, count the number of M&Ms in their bag, and compare that number with their initial guess. Then ask each child to sort the M&Ms by color and count the number of each color. Ask them how their predictions about the most and least abundant colors compared to the actual numbers.

One way for children to "tell the story" of a bag of M&Ms and its colors is to create a bar graph with their candies. Provide older children with large-square graph paper with a vertical line on the left and a horizontal line along the bottom. Students can list the M&M colors along the horizontal line and, along the left side, numbers zero to six, with zero at the bottom. Provide younger children with a guide for making their graphs; see Figure 3.1 on page 78. Children can then lay out the M&M's, by color, in vertical bars showing the amount of each color of M&M they found in their bag.

Integration to Other Content Areas

Reading Connections

Within the investigation, sorting and labeling items into various categories will help to generate descriptive oral vocabulary, which in turn supports the development of reading skills. During reading time, consider sharing books with children that describe sorting items by different attributes such as *Sorting by Size* by Jennifer Marks (2012) or *Sorting by Color* by Matt Bruning (2007) and then discuss with children things in your classroom that could also be sorted by that particular attribute. Encouraging children to interact with texts in meaningful ways and to apply what they discover in books to their own lives helps to build literacy and critical-thinking skills. There are also several engaging storybooks about sorting, including *Grandma's Button Box* by Linda Williams Aber (2002) and *Sort It Out* by Barbara Mariconda (2008).

Figure 3.1

Color Graphing

Color Counting	Blue	Brown	Green	Orange	Red	Yellow	
							6
							5
							4
							3
							2
							1

A full-size version of this figure is available at *www.nsta.org/startlifesci.*

Writing Connections

Create a class book about all the ways to sort seeds or beans. Let each child decide which attribute (size, shape, color, etc.) he or she wants to write about. Give each child a piece of paper, folded in half. At the top of each half, write for the child, or assist the child in writing, one of the two categories that he or she will use to sort the seeds or beans (e.g., big and small, bumpy and smooth, etc.). Children can then either draw the seeds or beans or glue them directly onto the appropriate side of their paper. Combine their work into a class book titled, "How We Sort Seeds" and share with the class, allowing each author to describe his or her page.

Math Connections

The concept of "sets" is an important one in mathematics. A set is a collection of objects considered as a whole. When children sort objects, they are creating sets (and often subsets as well). Early experiences with sorting and categorizing, together with your frequent use of the word *set*, can help prepare children for better future understanding of this aspect of mathematics. Provide children with sets of geometric shapes in easy-to-handle shapes, and invite children to sort the shapes into categories. Encourage them to count how many shapes are in each set or to identify which set has more shapes and then to find new ways of sorting the shapes. Remember that young children can typically sort by only one characteristic at a time.

Other Connections

Child's Life Connection

Not all seeds look like beans. Give each child a large sock to put on over one shoe. Go for a walk in a field. When you return, have children remove the socks and look for seeds that may be sticking to them. After removing the seeds, children can use magnifiers to closely examine these seeds and sort them based on their characteristics. Many of these seeds have coverings (i.e., fruits) that are "pokey" or have "stickers" or "burrs" on them, making them look very different from the seeds or beans examined earlier. Plants typically produce seeds in spring and summer, so this activity may be most effective in summer or early fall.

Center Connections

At the **art center**, provide children with an assortment of beans or seeds, some glue or paste, and a piece of paper that, if needed, has their name written on it for them. Children can build their fine motor skills as they write or trace their name in glue and then place seeds or beans on the letters to spell out their names. At the **sensory table**, have children participate in a blind sorting of seeds and buttons. Set up a touch box that children can reach into but not look into. In the box, place a bowl containing a mixture of seeds or beans and buttons from the science lesson. Have children sort seeds without looking. Children can reach into the middle bowl, select an item, and, using only touch, determine if it is a seed or a button. Placing two bowls, one for seeds or beans and one for buttons, outside of the box will allow children to place their selected item into the appropriate bowl before reaching into the box to select a new one—and helps make for easier cleanup. Ask children to sort the materials you have out in the **dramatic play** center. Have them decide on the strategy for sorting. They may choose to sort by any number of characteristics. If the thought of your dramatic play center sorted by color makes you a little anxious, you can always recommend that their final sort consider the materials' purpose within the center. "Put things that are part of our play kitchen together. Where should we put things that are used for dress-up?" By the end, your center might be the most organized it's been all year.

Family Activities

Your child has been studying sorting. An evening spent looking at some of your favorite family photographs can lead to a variety of interesting and challenging sorting activities. You might begin by sorting pictures into simple categories such as "girl" and "boy" pictures and then move on to others, such as sorting Mom's side of the family versus Dad's side or family members that live close by versus those that live far away. If you want to try something even closer to the way scientists classify things, try sorting into just two groups: one made up of pictures that display a certain characteristic and the other made up of pictures that do not display the characteristics. For example, you and your child could sort pictures into two categories: one called "cousins" and one called "not cousins." This is a great way for your child to practice sorting and to learn more about their family.

Actividades Familiares

Su hijo ha estado estudiando cómo clasificar. Pasar una noche viendo algunas de sus fotografías familiares favoritas puede llevar a una gama variedad de actividades de clasificación interesantes y desafiantes. Puede comenzar por ordenar las imágenes en categorías simples, como imágenes de "niña" y "niño" y luego continuar con otras, como clasificar por el lado de la familia materna versus el lado de la familia paterna o según los familiares que vivan más cerca versus los que viven más lejos. Si desea probar algo incluso más cercano a la forma como los científicos clasifican las cosas, intente ordenar solo en dos grupos, uno compuesto de fotografías que muestren ciertas características y el otro de imágenes que no muestran tales características. Por ejemplo, usted y su hijo pueden clasificar las imágenes en dos categorías, una llamada primos y otra llamada no primos. Esta es una forma estupenda para que su hijo practique clasificar y para aprender más sobre su familia.

Assessment—What to Look For

- **Can children describe seeds and other objects based on their observations?**
  (using senses to gather information, identifying properties, using complex patterns of speech)
- **Can children explain concepts of equality (more, less, equal; bigger, smaller, same size)?**
  (measuring, comparing size, using computational thinking)
- **Can children sort objects based on observable characteristics?**
  (identifying properties, comparing properties, classifying)
- **Can children explain their reasoning?**
  (using complex patterns of speech, constructing explanations, explaining based on evidence)

Standards

Head Start Early Learning Outcomes Framework
P-SCI 3. Child compares and categorizes observable phenomena. "Categorizes by sorting observable phenomena into groups based on attributes such as appearance, weight, function, ability, texture, odor, and sound."
Next Generation Science Standards
Science and Engineering Practice: Planning and carrying out investigations. "Make observations (firsthand or from media) and/or measurements to collect data that can be used to make comparisons."
Common Core State Standards for Mathematics
K.MD.B.3. Classify objects and count the number of objects in each category. "Classify objects into given categories; count the numbers of objects in each category and sort the categories by count."
Common Core State Standards for English Language Arts
L.1.5.A. Vocabulary acquisition and use. "Sort words into categories (e.g., colors, clothing) to gain a sense of the concepts the categories represent."

Seeds in Our Food

with Angelica Gunderson

Lesson: Comparing the seeds of different fruits

Learning Objectives: Children will use their observation skills to describe the characteristics of different types of seeds that each fruit has. Children will classify fruits and seeds in different ways.

Materials: Various fruits with different types of seeds (e.g., avocados, mango, melons, oranges, tomatoes, and apples); pictures of each fruit; writing and drawing materials; poster board or chart paper; plastic spoons; paper plates; magnifiers; paper towels; glue

Safety: Allergies to several different fruits and seeds are possible. Be sure to find out if any of your children have fruit or seed allergies before using these materials in your classroom and remind yourself of the symptoms of allergic reactions in children. Also, be aware of seeds as possible choking hazards.

Teacher Content Background: A common misconception about the fruits we eat is that the fruit serves as food for the seed, but this is not the case. Fruits are structures that protect and aid in the dispersal of seeds (i.e., the distribution of seeds away from the parent plant). For some plants, like the dandelion, fruits help the seed float in the breeze—yep, those bits of fluff are fruits, each carrying a small seed. Other fruits are adapted to be dispersed in water (e.g., coconut) or by temporarily attaching to an animal (the "burrs" that get entangled in your socks or poke into your bicycle tire are fruits). The fruits in this lesson also serve to aid in seed dispersal—in this case, by enticing animals to carry seeds, in their hands (picture a monkey and a mango) or in their bellies (picture yourself and a tomato), to new locations. Many whole seeds that are eaten are not digested and instead are "planted"—with a nice bit of fertilizer—after the seed passes safely through an animal's digestive system. Fruits develop from fertilized flowers. The reproduction of flowering plants is basically the same as our reproduction. Pollen contains sperm that is transferred to the flower's "eggs" (ovules) within the flower's ovary. Once fertilized, the ovules develop into seeds and the ovary becomes the fruit. (To learn more about flowers, see *Nature Bracelets*, p. 144.)

Science terms that may be helpful for teachers to know during this lesson include *investigate*, *sorting*, *seed*, and *fruit*.

Procedure

Getting Started

Introduction: Gather children around you and set out several different types of fruit for them to examine. [Note: Be sure to avoid seedless varieties of fruits.] Invite the children to pass the fruits around and to use their senses, including touch and smell, to make observations. Prompt them to share what they know and any questions they have about each fruit. Encourage children to identify the names of the fruits they know and share the names of unfamiliar fruits with children. Have children work together to arrange the fruits from largest in size to smallest. Then provide children with pictures of each fruit with the name of the fruit printed on it. Children can place these pictures in order by size on a poster, saying the name of each fruit as they do so. Then ask children for descriptions of each fruit from their observations to add to the poster beneath each picture.

Initial Explanation: Ask children, "These pictures show the outside of the fruit, but what about the inside? What do you think the fruit looks like on the inside?" After children have discussed their initial ideas, assign each student, possibly in partners or in small groups, a fruit. Give children paper plates and crayons. Have children draw their ideas about the inside of their fruit on the plate. Children will use their drawings later to compare their initial ideas with observations of the actual fruits.

Investigating

Observing: Explain to children that they will now investigate the inside of the fruit. [Note: In an area away from children, precut each fruit in half, but provide a whole fruit (i.e., both halves) to each team of children.] Allow children to explore the inside of the fruit as they wish. Provide tools, such as magnifiers and plastic spoons, to assist children. Use questioning to encourage children to look closely at fruits, but more importantly, *listen* to your children—follow their leads, note their discoveries, and let them tell you about their fruits and the seeds within.

Comparing: After children have explored their fruits in ways meaningful to them, direct their attention to the seeds. Help children to remove the seed or seeds from their fruit. Have children use paper towels to clean the seeds and place them on their paper plates. Provide children with magnifiers and prompt them to make closer observations of the seeds. Encourage children to compare their seeds to their initial drawings. Natural comparisons here will be the number of seeds and the size of seeds.

After comparing the insides of their own fruits to their initial ideas, provide trays or bins and let children set up a station that shows what they learned about their fruit. On their tray, children can place their initial picture, the seeds they collected, a whole or sliced fruit (or a picture of the fruit), and a piece of paper that they can write the fruit name on or that has the fruit name printed on it for them. Children can then go on a "gallery walk," visiting each other's stations and exploring the seeds of different fruits. It's helpful if children complete the gallery walk in groups of two or three so that they can describe and discuss the fruits and seeds they see.

Making Sense

Identifying Patterns: Ask children to share their most surprising discovery about the fruits. Let them discuss findings of interest to them. Then direct the conversation toward what they found inside their fruits and toward the defining feature of fruits: All fruits have seeds. Ask children, "What's one thing that all fruits have?" With younger children, use a questioning sequence that helps guide them to the conclusion that all fruits have seeds. For example, "Did the avocado have a seed? What about the apple; did the apple have seeds? Did all of the fruits have seeds?"

Describing Findings: Revisit the chart made based on fruit size at the beginning of the lesson. Add to this chart the information children share about the seeds and then encourage children to discuss the findings. Ask questions such as "What do you notice about the seeds? Do all the fruits have the same number of seeds? Are all seeds the same size? Do all seeds feel the same? Do all big fruits have big seeds? Do all small fruits have small seeds?" Encourage children to use their observations to explain or justify their responses. For example, if a student states that fruits don't all have the same number of seeds, ask him or her to tell you two fruits that have a different number of seeds. Help your children to understand that different types of fruits have different numbers, types, and sizes of seeds.

Investigating fruit

What's Next?

Extension Activity

Prompt children to think about how fruits may help plants, asking, "How do fruits help seeds spread to new areas?" Share with the children a book that describes how fruits help seeds to move. Several books describe this important function of fruits and seeds, including *Flip, Float, Fly* by JoAnn Early Macken (2008), *What Kind of Seeds Are These?* by Heidi Bee Roemer (2006), *Who Will Plant a Tree?* by Jerry Pallotta (2010), *From Bird Poop to Wind: How Seeds Get Around* by Ellen Lawrence (2012), *Seeds* by Ken Robbins (2005), *Planting the Wild Garden* by Kathryn Galbraith (2015), *Next Time You See a Maple Seed* by Emily Morgan (2014), and *A Fruit Is a Suitcase for Seeds* by Jean Richards (2006). After children have read about the different ways seeds are moved, they can test different fruits. Share with children several of the fruits or seeds described in the books. For each seed, ask children, "How can we test how this seed moves?" Use a fan, a water table, and fake fur or feathers and assist children in testing the different ways to move fruits and seeds. If you'd like you can structure these tests as formal science investigations, having children collect and analyze data to use in supporting their claims about how different seeds move.

Integration to Other Content Areas

Reading Connections

There are many opportunities for children to engage with texts to learn more and to think creatively about fruits and seeds. Examples of books that describe seed dispersal are provided in the *Extension Activity* section above. Additionally, Vijaya Khisty Bodach has written a Plant Parts Series and each book shows several examples of a particular plant part and describes their uses; the books, *Seeds* (2016) and *Fruits* (2007) can help children to see examples of the fruits or seeds they explored in class as well as additional ones they have yet to explore. Children can use the pictures in these books or others such as *Why Do Plants Have Fruits?* by Celeste Bishop (2016), *Flower to Fruit* by Richard Konicek-Moran and Kathleen Konicek-Moran (2017), *Fruits* by Julia Adams, and *Fruits* by Grace Hansen (2011) to "read" as they describe what they learned about fruits during the lesson. Enjoyable narrative books such as *Handa's Surprise* by Eileen Browne (1999) and *Fruits: A Caribbean Counting Poem* by Valarie Bloom (1997) can also share additional examples of fruits. And more text-rich books, such as *Seeds and Fruit* by Melanie Waldron (2014) and *The Fruits We Eat* by Gail Gibbons (2016), can provide children with even more science information about fruits and their seeds.

Writing Connections

Have children draw, or take and print photographs of, several of the fruits that you studied during your lesson. Make sure there is space on the page for children to write the word *Fruits* at the top and to label each of the fruits. Ask children to write the names of the fruits. As children write, encourage them to use the letters of the most conspicuous sounds in the words. Help children identify and write the first letters of the fruits by emphasizing the beginning sound of each fruit. "This is a mmmmango. What's the first sound you hear in the word, *mmmmango*? Good. So, as you write mango, which letter should you write first?" Older children can use their labeled page as the basis for constructing sentences. Having written the word *fruits* and the names of two or more fruits, children can construct sentences such as "Apples and figs are fruits." Provide a sentence frame such as, "___ and ___ are fruits" to help children to structure their writing.

Math Connections

This lesson presents several opportunities for children to represent and interpret data through counting and graphing. Provide children with fruits that have different numbers of seeds and a graphic organizer, like the one shown in Figure 3.2, to help children to organize their data. Slice these fruits for children and allow them to remove and count the seeds they find. Avoid fruits with hard flesh, such as apples. Instead opt for fruits with easily excavated seeds such as peaches, oranges, or grapes. You can also provide children a fruit that has a number of seeds appropriate for the numeracy development of your children—pumpkins or papayas might be perfect for some teams of children but might have too many seeds for others. As children count seeds, encourage them to arrange seeds in sets of two, or three, or five, as appropriate for each child. Children can represent the numbers of seeds by creating a bar graph that shows how many seeds were found in each fruit. Children can also graph the seeds according to other similarities the seeds have, such as size, shape, or color.

Figure 3.2

Color Counting

One Seed	*A Few Seeds*	*Many Seeds*
1		

A full-size version of this figure is available at *www.nsta.org/startlifesci.*

Other Connections

Child's Life Connection

Find an assortment of fruits and vegetables that are familiar to children. If your school serves lunch to the children, see if your cafeteria staff can provide you with a menu for upcoming meals. Perhaps they can even supply you with samples and perhaps knowing that your class is studying fruits will inspire them to add a greater variety of fruits to the menu. Children can explore the vegetables and fruits, sorting the produce into two groups: foods that have seeds (fruits) and foods that don't have seeds (vegetables). Challenge your children by adding fruits and vegetables that, although fairly common and available at your local market, may be less familiar to them (figs, kiwi, bok choy, parsnips, etc.). (See *Where Vegetables Come From*, p. 88, to expand on this follow-up activity.)

Center Connections

Faces on fruit shouldn't just occur on pumpkins in October. In the **art center**, children can paint "fruit faces" on a variety of different fruits. Let each child select a fruit and then provide paints; paste; yarn; precut eyes, noses, mouths, and ears; and anything else needed for children to create a fruit face (e.g., hats, mustaches, etc.). At the **sensory table**, put different seeds from the lesson in a touch box for children to feel. Encourage children to guess which seeds are which. Provide children with the sentence structure to help develop their oral language skills as they describe their reasoning. "I think this is a mango seed because it is big. I think this is a watermelon seed because it is flat. I think this is an avocado seed because it's big and round." Provide and encourage children to use spoons, tweezers, and other tools to manipulate the various-sized seeds. In the **dramatic play** center, children can use fruits to put on a puppet show. Provide different fruits and materials so that children can transform the fruits into faces (see the Art Center activity). Once the fruits have become "people," children can use them as they would puppets to talk to each other and to act out different scenarios that they imagine and negotiate.

Family Activities

Your child has been learning about seeds and fruits. Fruits grow from flowers and contain seeds. Many of the foods we commonly call vegetables (eggplant, tomato, cucumber, etc.) are actually fruits. With your child, investigate the different foods your family eats. As you are preparing dinner, ask your child if she or he thinks a certain food has seeds or not and then cut it open to find out. Celebrate your discoveries with your child. "It has seeds! It must be a fruit." Take your child to the produce section of your local market and purchase new foods to learn about. Back at home, search for seeds to determine if the new item is a fruit or not and then prepare and sample the new food together. As you and your child learn science together, you may discover a new favorite food to share!

Actividades Familiares

Su hijo ha estado aprendiendo sobre semillas y frutas. Las frutas crecen a partir de las flores y contienen semillas. Muchos de los alimentos que por lo general llamamos vegetales (berenjena, tomate, pepino, etc.) de hecho son frutas. Con su hijo, investigue los distintos alimentos que come su familia. A la hora de preparar la cena, pregúntele a su hijo si cree que un alimento en particular tiene unas semillas y luego ábralo para descubrir. Celebre los descubrimientos con su hijo: "¡Tiene semillas! Debe ser fruta". Lleve a su hijo a la sección de alimentos en su mercado local y compre comidas nuevas sobre las cuales aprender. Al llegar a casa, busque si tienen semillas para determinar si el producto es o no fruta, y luego preparen una muestra de la nueva comida juntos. A medida que usted y su hijo aprenden sobre ciencia juntos, ¡puede descubrir una nueva comida para compartir juntos!

Assessment—What to Look For

- **Can children describe characteristics of seeds based on their observations?**
  (using senses to gather information, identifying properties, using complex patterns of speech)
- **Can children sort seeds based on observable characteristics?**
  (identifying properties, comparing properties, classifying)
- **Can children infer the purpose of seeds and their structures?**
  (making inferences, identifying causality, using complex patterns of speech)
- **Can children record (draw or write) observations and contribute to class discussion?**
  (using new or complex vocabulary, documenting and reporting findings, discussing scientific concepts, listening to and understanding speech)

Standards

Head Start Early Learning Outcomes Framework
P-SCI 1. Child observes and describes observable phenomena (objects, materials, organisms, events, etc.). "Makes increasingly complex observations of objects, materials, organisms, and events. Provides greater detail in descriptions. Represents observable phenomena in more complex ways, such as pictures that include more detail."
Next Generation Science Standards
Science and Engineering Practice: Constructing explanations and designing solutions. "Make observations (firsthand or from media) to construct an evidence-based account for natural phenomena."
Common Core State Standards for Mathematics
1.MD.A.1. Measure lengths indirectly and by iterating length units. "*Order three objects by length; compare the lengths of two objects indirectly by using a third object.*"
Common Core State Standards for English Language Arts
L.1.5.B. Vocabulary acquisition and use. "*Define words by category and by one or more key attributes (e.g., a duck is a bird that swims; a tiger is a large cat with stripes).*"

Where Vegetables Come From

with Angelica Gunderson

Lesson: Identifying the parts of plants that different vegetables come from

Learning Objectives: Children will make observations and comparisons of vegetables to understand that each comes from a part of a plant.

Materials: Various vegetables that come from different parts of plants such as roots (carrots), leaves (spinach), stems (asparagus), flowers (broccoli) and fruits (tomatoes); [Note: You may also choose to include leaf stalks or "petioles," such as celery, in this lesson]; plastic spoons; paper plates; magnifiers

Safety: Allergies to several different fruits and vegetables are possible. Be sure to find out if any of your children have food allergies before using fruits and vegetables in your classroom and remind yourself of the symptoms of allergic reactions in children. Also, be aware of seeds as possible choking hazards.

Teacher Content Background: Vegetables are foods that come from different parts of a plant, including leaves (lettuce, spinach); roots (carrots, turnips); stems (asparagus, ginger); fruits (zucchini, eggplant); and even flowers (broccoli, artichokes). Along with fruits, vegetables are important parts of a healthy diet, as they provide important vitamins and minerals. Federal dietary guidelines recommend five to nine servings of fruits or vegetables per day, representing a third to one half of the food you eat. A diet with a variety of different fruits and vegetables is important, as different foods provide different vitamins and minerals. Generally, fruits and vegetables that are yellow, orange, or red in color are rich in vitamins A, B_5, B_6, B_7, and C, as well as the mineral potassium, whereas dark-green vegetables are rich in vitamins B_1, B_6, B_9, E, and K, and many minerals including calcium, iron, and magnesium. Encourage children to eat many different colors of food each day, helping them to achieve a healthy and nutritious diet.

Science terms that may be helpful for teachers to know during this lesson include *comparing, sorting, vegetable, root, stem, leaf, flower,* and *fruit.*

Procedure

[Note: In this lesson, children explore and compare different vegetables. Using unfamiliar vegetables will increase children's curiosity and lead to more critical thinking and better investigations.]

Getting Started

Introduction: Place a number of different potted plants in front of the children and encourage them to observe the plants. Have a conversation with children about the similarities and differences among the plants. Welcome and listen to children's ideas before guiding them to focus on plant parts. Encourage children to identify and compare different plant parts by asking questions such as "Do all of these plants have leaves? How do all of these leaves look?" Repeat this type of questioning for each plant part: stems, roots, flowers, and, if present, fruits. Summarize the conversation by telling children that plants have five parts and ask children to name the parts. Use a diagram or other visual to help show the typical location and structure of each part.

Investigating

Observing and Comparing: Introduce children to an assortment of vegetables. Let children tell you the names of the vegetables they know and then review and tell them the name of each food. Prompt children to tell you what they know about each vegetable, and encourage them to share any questions they have. Encourage children to observe the vegetables, allowing children to hold and feel each of them. Provide tools, such as magnifiers and plastic spoons, to assist children as they explore each vegetable. (If children are examining hard fruit and vegetables, you may want to slice them for children so they can observe the insides as well.) Use questioning to encourage children to look closely at each vegetable and to share their findings with their peers. Engage children in a conversation about the vegetables, encouraging them to describe the ways some vegetables are similar and also the ways some are different.

Sorting: Let children know that each of the vegetables is part of a plant, and that they will try to figure out which plant part each vegetable is. Set out five bins, each labeled with the name of a different plant part (leaf, stem, root, flower, fruit). Review each of these words and plant parts with children as you put out each bin. Then have children collaborate to decide how to sort each vegetable. [Note: Younger children will need a vegetable or a set of vegetables of their own to sort.] As children sort each vegetable, engage them in conversation, encouraging children to explain the reasoning behind their decisions.

Making Sense

Describing Findings: Review the results of children's investigations, one plant part at a time. Ask children, "Which vegetables do you think are roots?" and then, "How can you tell if a vegetable is a root?" Help to structure children's responses by providing prompts such as "I think ____ is a root because ____." After discussing roots, move on to the next plant part and continue the conversation, reviewing all five plant parts. Throughout the conversation, don't worry if children's responses are right or wrong. Instead, emphasize children's reasoning, making sure children can offer a justification that is based on their observations. Anticipate responses such as, "I think the spinach is a leaf because it is green" or "I think the bell pepper is a fruit because it has seeds." If possible, have images of the plants each vegetable comes from. Let children examine these images to help them make sense of which plant part each vegetable comes from. If children can't agree on the part a particular vegetable comes from, consider calling in a vegetable expert, perhaps from the school cafeteria, to help explain.

Application: Bring in new, different vegetables. Introduce each to the class and then let children observe these new vegetables. As they explore, ask children, "What part of a plant does this vegetable come from? How do you know?" After these observations are complete, review each vegetable as a class. This type of application is important, as it clarifies and reinforces what children have learned.

What's Next?

Extension Activity

Plant a vegetable garden. Raised planter beds are often the best option; they can be made to fit any space and can be installed on top of soil or pavement. However, if these structures aren't feasible for your school, consider container gardens. Although generally the bigger the better (as roots will need room to grow), any size container will work. Consider old buckets, washtubs, and so on. Because plants in these containers will have limited amounts of soil, you may need to water them more frequently. Once your garden beds or containers are built and filled with soil, you're ready to plant seeds and start growing. Be sure to involve children in this entire process —planning, building, filling, planting, and tending the garden. If an outdoor vegetable garden is not possible at your school site, plant a small indoor herb garden instead. Plant cilantro, chives, dill, parsley, and mint in small pots in an area that gets indirect sunlight and water the pots regularly. (See *Sprouting Seeds*, p. 94, for information about germination and consider completing the *Sprouting Seeds* lesson before this extension activity.)

Integration to Other Content Areas

Reading Connections

There are many great children's books about fruits and vegetables. For books about fruits, check out *Seeds in Our Food* on page 81. To help children to learn more about the great variety of vegetables, share books like *Rah, Rah, Radishes* by April Pulley Sayre (2014); *The Vegetables We Eat* by Gail Gibbons (2008); and *Up, Down, and Around* by Katherine Ayers (2008). There are also several books that can inform and inspire your children to eat healthy, well-balanced diets, such as *Good Enough to Eat* by Lizzie Rockwell (2009), *My Plate and You* by Gillia Olson (2011), and Nancy Dickmann's *Vegetables* (2012). And there are many entertaining books about eating and trying new foods, such as *Oliver's Vegetables* and *Oliver's Fruit Salad* by Vivian French (1995; 1998), *Monsters Don't Eat Broccoli* by Barbara Jean Hicks (2014), *I Will Never Not Ever Eat a Tomato* by Lauren Child (2003), and *The Seven Silly Eaters* by Mary Ann Hoberman (2000) that children and you will enjoy.

Writing Connections

Write the alphabet on a long strip of paper and have children place fruits and vegetables on the letter that matches the letter that begins the name of each. From *asparagus* to *zucchini*, there are a great many possibilities. After children have sorted the fruits and vegetables by their beginning letter, read with the children an alphabet book of fruits and vegetables, such as *An Alphabet Salad* by Sarah Schuette (2003) and *Eating the Alphabet* by Lois Ehlert (1996). As you read, ask children to recall which fruits or vegetables they used for each letter. Use this experience to inspire your class to make its own alphabet book of fruits and vegetables. Take a picture of each child holding a fruit or vegetable and a card with its name. [Note: If you photograph children holding fruits or vegetables that begin with the same letter as the child's name (e.g., Brenda holding broccoli, Victor holding a Valencia orange, etc.), you may be able to avoid having to track down a quince or a ximenia.] Print the pictures and have children arrange them in alphabetical order for their class book. As children read the book, encourage them to identify each letter and letter sound.

Math Connections

Different fruits and vegetables come in all different sizes and shapes. This variety is a perfect opportunity for children to make measurements and size comparisons. Children can arrange fruits and vegetables in order based on length, width, and weight. For length and width, children can use rulers to measure each fruit and vegetable, or they can place yarn along (for length) or across (for width) the fruits and vegetables and cut the yarn into pieces that represent the distance measured. Children can compare these measurements and arrange fruits and vegetables in order based on their size. For weight, have children sequence fruits and vegetables based on how heavy they feel, then provide children with a balance. Show children how to compare the weight of two items using the balance and encourage them to use comparative language such as, *heavier than* or *lighter than* as they describe their findings.

Other Connections

Child's Life Connection

Healthy eating is an important part of all children's lives. Use your study of fruits and vegetables to discuss and inspire healthy eating. Share books such as *Fruits on My Plate* or *Vegetables on My Plate* both by Mari Schuh (2012; 2012) or *Good Enough to Eat* by Lizzie Rockwell (1999) and engage children in conversations about the importance of fruits and vegetables in a healthy diet. If your school serves lunch, have children collaborate to keep track of the number and kinds of fruits and vegetables served. Establish a bar graph and add to it each day after lunch. Engage children in discussions about the different foods they've eaten and of any confusion they might encounter (where are the seeds in apple juice, seedless oranges, blueberries, etc.). Be ready to follow up and let children explore any fruits or vegetables that need further investigation. Maybe you can get a "nutrition expert" from your school cafeteria to visit your class. During your conversation, children can share their graph with your guest, and he or she can encourage children to eat healthy at school and at home.

Center Connections

At the **art center**, provide children with paper with the word, *vegetable* printed on it and then have children use different vegetables as paint brushes to paint

pictures of five different plant parts: one each of roots, stems, leaves, flowers, and fruits. Encourage children to select leaf "brushes" for their leaf painting, fruit ones for their fruit painting, and so on. At the **sensory table**, set up a touch box that children can reach in but not look into. In the box, add an assortment of fruits and vegetables. Have children focus on and describe the texture of different fruits and vegetables. Challenge children to sort the vegetables in different ways (e.g., fruits and flowers to one side of the box and the roots, stems, and leaves to the other). For **dramatic play**, children can run a produce stand or farmer's market, buying and selling different fruits and vegetables. Provide items such as plastic (or real) fruits and vegetables, paper bags, small shopping carts or baskets, a cash register, aprons, pretend money, crates or boxes, and materials for children to make signs for the different fruits and their prices. Children can make shopping lists and pay for their items. Encourage children to use multiple senses to describe the produce they purchase.

Family Activities

Your child has been learning about different types of vegetables. Vegetables are parts of plants that we can eat; all vegetables are leaves, roots, stems, flowers, or fruits. With your child, investigate the different foods your family eats; ask your child what part of the plant he or she thinks each vegetable comes from. How many of the vegetables in your refrigerator or pantry are leaves, roots, stems, flowers, or fruits? Together with your child, categorize and count the vegetables in your home. Take your child to the produce section of your local market and have your child identify different types of vegetables. Ask your child what part of the plant each comes from. Go on a scavenger hunt together, searching for vegetables that are roots. Enthusiastically, ask your child, "I wonder how many different roots we can find?" and search your market together. Repeat your hunt for stems, leaves, flowers, and fruits, having fun with your child as you explore the different plants we eat.

Actividades Familiares

Su hijo ha estado aprendiendo sobre los distintos tipos de verduras. Las verduras son partes de las plantas que podemos comer; todas las verduras son hojas, raíces, tallos, flores o frutas. Con su hijo, investigue los distintos alimentos que come su familia; pregúntele a su hijo de qué parte de la planta él o ella piensa que proviene cada vegetal. ¿Cuántas de las verduras en suel refrigerador o la despensa son hojas, raíces, tallos, flores o frutas? Junto a su hijo, clasifique y cuente las verduras en su casa. Lleve a su hijo a la sección de alimentos de su mercado local y pídale que identifique los diferentes tipos de verduras. Pregúntele a su hijo de qué parte de la planta proviene cada uno. Embárquese en una búsqueda del tesoro en conjunto, en busca de verduras que son raíces. Con entusiasmo, pregúntele a su hijo: "Me pregunto, ¿cuántas raíces distintas podemos encontrar?" y busquen juntos en el mercado. Repita su búsqueda de tallos, hojas, flores y frutos, divirtiéndose con su hijo mientras explora las diferentes plantas que comemos.

Assessment—What to Look For

- **Can children make predictions about the properties of fruits and vegetables?**
  (making predictions, making inferences, using complex patterns of speech)
- **Can children describe where seeds are found based on their investigation?**
  (identifying properties, using new or complex vocabulary)
- **Can children compare their predictions to their actual findings?**
  (comparing properties, explaining based on evidence, using new or complex vocabulary)
- **Can children record (draw or write) observations and contribute to class discussion?**
  (using new or complex vocabulary, documenting and reporting findings, discussing scientific concepts, listening to and understanding speech)

Standards

Head Start Early Learning Outcomes Framework
P-SCI 2. Child engages in scientific talk. "Uses scientific content words when investigating and describing observable phenomena, such as parts of a plant, animal, or object."
Next Generation Science Standards
Science and Engineering Practice: Planning and carrying out investigations. "Make observations (firsthand or from media) and/or measurements to collect data that can be used to make comparisons. Make predictions based on prior experiences."
Common Core State Standards for Mathematics
K.MD.B.3. Classify objects and count the number of objects in each category. "Classify objects into given categories; count the numbers of objects in each category and sort the categories by count."
Common Core State Standards for English Language Arts
L.1.5.B. Vocabulary acquisition and use. "Define words by category and by one or more key attributes (e.g., a duck is a bird that swims; a tiger is a large cat with stripes)."

Sprouting Seeds

with Lauren M. Shea

Lesson: Observing lima bean seeds as they germinate

Learning Goals: Children will observe and describe the physical properties of lima bean seeds and observe how the characteristics change when the seeds are soaked in water. They will observe and compare the germination of soaked and unsoaked seeds.

Materials: A package of lima bean seeds (available at most grocery stores), writing and drawing materials, magnifiers, clear plastic cups, water, small, clear plastic bags, paper towels, tape, potting soil

Safety: Be aware of beans and other seeds as potential choking hazards. Remind children not to eat any food being used in the activity. Only use pesticide-free and sterilized potting soil and be sure to instruct children to not put their fingers in their mouth or nose while handling potting soil. Make sure any spilled water is wiped up to prevent slipping hazards. Remind children to keep away from electrical outlets when working with water. Make sure children thoroughly wash their hands with soap and water after the activity.

Teacher Content Background: Seeds are small, dormant, embryonic plants waiting for the right environmental conditions to begin growing. The growth of a plant from a seed is called germination. To germinate, all seeds require water, oxygen, and appropriate temperatures—although many species of plants also have additional, and often very specific, requirements (soil type, light levels, etc.). The uptake of water by seeds, a process called imbibition, initiates germination. The imbibition of water softens tissues within the seed and causes the seed to swell. As the seed enlarges, the softened seed coat splits open. Within the seed, water triggers a series of chemical changes that activate seed metabolism, resulting in energy for the embryonic plant within the seed. As the embryo grows, the root emerges from the seed and, sensing gravity, grows downward. The shoot then emerges, growing toward light. The once-dormant seed has germinated and is transformed into an active, growing seedling.

Science terms that may be helpful for teachers to know during this lesson include *observe, seed, seedling, germinate,* and *seed coat.*

Procedure

[Note: Although lima beans are used here, any other fast-sprouting seed can be used, such as corn, soybeans, and most other types of beans. The advantage of lima beans their its large size, making them easier for children to manipulate and observe.]

Getting Started

Prior Knowledge: Ask children to think about seeds and to share with the class what they know about seeds, including where seeds are found and what seeds do. Also, encourage children to ask any questions they have about seeds. [Note: Avoid the temptation to answer these questions. Instead, acknowledge and validate each question and let children know that they're going to investigate seeds and that they may discover some of these answers on their own.] Then direct children toward observable characteristics of seeds, asking, "If we were to observe a seed, what would we notice?" Review the five senses and prompt children to consider what each sense could tell us about seeds (e.g., "What do seeds feel like? What shapes are they? What sizes? Colors?"). (See *Scavenger Hunt*, p. 177, for a discussion about using the senses to make observations.)

Investigating 1

Observing: Give each child a lima bean seed and a magnifier. Prompt children to observe the seed, asking questions such as "What does the seed look like? How does it feel?" [Note: Remind children that they will not be using their sense of taste to make observations.] Tell children that they're going to soak the seeds in water overnight. Ask them to predict how soaking might change the seeds. Listen to the children's predictions and encourage them to explain their reasoning. Let children help you place seeds (enough for three per child) in water to soak. Throughout the rest of the day, allow children to check the seeds to see if they are changing.

Comparing: The next day, take the seeds out of the water and give each child one to observe. Have the children look closely at their seed and discuss their findings with a partner. Prompt children to think about how the seed has changed while it soaked in the water. "Does the size, texture, shape, or firmness of the bean seed seem different today?" Provide unsoaked bean seeds for comparison. As children compare the bean seeds, encourage them to describe differences between the soaked and unsoaked seeds.

Investigating 2

Experimenting: After observing seeds, it's time to test to see if the seeds can germinate and become plants. Describe and demonstrate for children how they will "plant" their seeds, wrapping two unsoaked beans in a damp paper towel, placing the towel into a clear plastic bag, and sealing the bag part way, leaving about a 2-inch opening and then repeating the process for two soaked beans. Ask the children what they think will happen to the soaked and the unsoaked seeds; discuss and record their ideas. Then provide children with materials and assistance needed to "plant" their seeds. Children can label each bag with their name and the words *soaked* and *not soaked*. Tape the bags to a window inside the classroom. [Note: Sunlight is important for the seedling, but make sure the window you select does not get direct sunlight, as high temperatures can affect germination and growth. Each day, use a spray bottle set to mist to keep the towels moist, but be sure not to overwater the seeds, as this may cause the bean to rot and mold to form. (Avoid having water sitting at the bottom of the bag.) Soaked seeds will germinate in about a week, unsoaked seeds will take about twice as long, and a few seeds, soaked or not, may not germinate at all.] As the seeds germinate, roots will appear first, followed shortly by the emergence of a green shoot. Watch for these developments; once they occur, the seed has germinated. Children can photograph the changes they see, count the number of germinated soaked and unsoaked seeds each day, and record their data (see Figure 3.3).

Figure 3.3

Germination Data Table

Number of Germinated Seeds		
Day	Soaked	Not Soaked
1		
2		
3		
4		
5		
6		
7		
8		
9		
10		
11		
12		
13		
14		

A full-size version of this figure is available at *www.nsta.org/startlifesci.*

Making Sense

Describing Findings: Have children review the germination data collected during the *Investigating 2* activity. To help children to better understand the data, you may choose to create two bar graphs: one for soaked seeds and one for unsoaked seeds. Along the bottom of each graph, put the number of days, and each day plot the number of seeds that have germinated. Typical data will show that soaked seeds germinate sooner than unsoaked seeds and that more of the soaked seeds germinated overall. Encourage children to share their ideas about why more soaked seeds germinated and why the soaked seeds germinated sooner.

Identifying Patterns: Use pictures to review the seed-growing process and encourage children to identify the steps that took place for the seed to grow (e.g., seeds take in water and swell, the seed coat splits open, roots appear, and then the stem and leaves emerge). Encourage children to share their ideas about and understanding of the process, asking questions such as, "What did the seeds need to be able to germinate? Which part of germination do you think was most interesting? Why did you like that part? What questions about germination do you have?" See if children can generalize their knowledge to other seeds that they have learned about, asking, "Do you think all seeds germinate this way? Do you think peach or cherry seeds need water and sunlight to grow?" This conversation will help to expand children's curiosity about seeds and may lead to all sorts of new explorations.

What's Next?

Extension Activity 1

Children can dissect a (soaked) seed and use magnifiers to investigate the different parts of a seed. [Note: Many children will need support with this investigation. Also. If *dissect* is a new word for your class, take time to support children's understanding of this new term—reminding them that to dissect means to take something apart very carefully so you can observe its individual parts.] Model for children (and assist as needed) how to carefully remove the seed coat of the seed. Once it is removed, encourage children to think about and observe the seed coat, asking questions such as "How does the seed coat look? Why do you think a seed has a seed coat?" Then show children how to open the seed at the seam between the two halves. These halves are the endosperm that provide food for the seedling during germination, and between them is an embryo. Children are often excited to discover this "baby plant." Encourage children to describe the embryo and explain what they think it might be.

Extension Activity 2

Once seeds have germinated, have children transfer their seedlings into pots with soil. For several days, children can make daily observations and measurements of their growing seedlings. Encourage children to talk about what they notice happening to the seedlings. Every few days, children can draw life-sized pictures that record details of the seedlings' growth (labeled "Day 3," "Day 6," etc.). Encourage children to count the seedlings' leaves and to include the same number of leaves in their drawings. [Note: The data tables included in *Terrariums*, p. 101, may be helpful here.] With your help, children can lay the seedlings along their paper to make sure their drawing accurately represents the seedling's height. As new details or plant parts appear, give children vocabulary (e.g., *stem, cotyledon, leaf*) for each part and help children to label these on their drawings. Comparing their drawings will help children gain an appreciation for the growth of their seedling.

Integration to Other Content Areas

Reading Connections

The activities in this lesson include opportunities to draw children's attention to print and to use print in meaningful ways that support children's development of emergent literacy skills. These include describing and drawing seedlings, measuring and documenting growth, and labeling drawings. Expand children's use of print by filling your classroom library with both fiction and nonfiction books about the sprouting of seeds. Some possible books to share include *One Bean* by Anne Rockwell (1999), *Seeds! Seeds! Seeds!* by Nancy Elizabeth Wallace (2013), *How a Seed Grows* by Helene J. Jordan (2015), *From Seed to Plant* by Gail Gibbons (1991), *A Seed Grows* by Pamela Hickman (1997), and even a version of the *Jack and the Beanstalk* fable. As you read books with the class, have conversations with children about their science investigations to determine if different aspects of the books are factual or fictional. Ask children questions such as "Do plants reallly grow like this? Is this how seeds germinate? Can this really happen? How is this similar to what happened with our seeds?"

Writing Connections

As children draw, label, and write about their investigations, they engage in writing as a form of scientific communication that allows for documentation of thinking, ideas, and observations. Children can further develop their writing skills, as well as their fine motor skills, using beans. Provide children with large, printed letters or words and prompt children to write or trace these letters or words with beans. Children can also develop their sequencing skills, important for both language use and science. Provide pictures of bean seeds in different stages of germination; perhaps use the photographs taken by your children during the *Investigating 2* activity. Children can label different structures (e.g., seed coat, stem, leaves, roots), sequence the pictures, and describe the germination process to each other. Provide children with language supports such as "The first step was ___. The next step was ___. The final step was ___."

Math Connections

Plant growth provides great opportunities for measurement and for graphing. As each seed sprouts, children can use and develop measuring skills as they determine the length of the seedling. Guide children to use a ruler to take daily measurements in inches or centimeters. With your support, children can assist each other in taking and recording these standard measurements. Prompt children to compare the size of their seedling to objects around the room (e.g., "My plant is longer than a crayon"). After several days of measurement, children can create graphs showing how the lengths of their seedlings have changed from day to day. Set up the graphs with the day or date on the bottom (horizontal axis) of the graph and the seedling length on the side (vertical axis) of the graph, making sure the size of your graph corresponds to the actual lengths of seedlings. Then help children to plot their data, drawing seedlings of increasing length for each day.

Other Connections

Child's Life Connection

Children encounter many different seeds in their daily lives, especially in and as food. Have children find and collect seeds at home. Encourage children to think about what they could do to help these seeds to germinate. With help at school or home, children can conduct investigations with these seeds similar to the lesson with lima beans—soaking the seeds in water to see if they have seed coats that can peel off easily and placing them in bags with wet paper towels to see if they'll germinate. Children can compare their results and discuss which seeds germinated and which did not. [Note: Not all seeds found in foods will germinate, mostly because these seeds are sterile due to breeding or processing or because the seeds have specific needs (in addition to water) that have not been met.]

Center Connections

Because different types of beans are easy to obtain and play with, they can be used in many centers throughout the classroom. At the **art center**, children can develop their fine motor skills as they create "bean art." Provide children with paper, glue, paintbrushes, and beans of various sizes and colors. Children can brush on the glue in any design they wish, or perhaps a picture of a seedling, and then add beans to complete their art. Fill half of the **sensory table** with assorted beans with different colors, sizes, shapes, and textures for children to touch, dig, and play with. Add cups, tubes, large funnels, and other tools. Encourage children to use words to describe how the beans feel and sound. On the other half of the sensory table, put the same assortment of beans, but add water. Children can feel how soft the beans get, see the seed coats peel off, and hear the difference in their movement. In the **dramatic play** area, children can be gardeners and pretend to grow plants from seeds. Plastic gardening tools, child-sized gardening gloves, aprons, empty seed packets, watering cans, pots, seeds, and (artificial) plants can create an environment where children can explore how to take care of a garden and help plants grow.

Family Activities

In school, your child has been investigating lima bean seeds. We soaked the seeds in water for 24 hours and then observed how the seeds changed. After soaking, we were able to remove the outer layer, called the seed coat, and split the seed in half. When the seed was opened, we observed the inside of the seed to identify the baby plant and its food inside. With other seeds, we watched how a lima bean seed can germinate *and grow into a plant. To do this, we first soaked the beans overnight and then wrapped them in a damp paper towel, placed the seeds and towel in a clear plastic bag, and taped the bag to a window. At home, you can help foster your child's learning by repeating our investigations together. Let your child take the lead and show you how to soak the beans and how to identify the different parts of the seed. Demonstrate your own interest and curiosity by asking questions such as "I wonder what the inside of a lima bean looks like—do you know? What happens when you soak the beans in water? How can we help the seed to grow?" Together, germinate the beans in a clear plastic bag. After they sprout, you and your child can plant them in a small pot with soil and watch them grow together.*

Actividades Familiares

En la escuela, su hijo ha estado investigando con semillas de frijol de Lima. Remojamos las semillas en agua durante 24 horas y luego observamos cómo cambian las semillas. Después del remojo, pudimos eliminar la capa exterior, llamada tegumento, *y dividir la semilla por la mitad. Cuando se abrió la semilla, observamos el interior para identificar la planta del bebé y su comida en el interior. Con otras semillas, observamos cómo puede* germinar *una semilla de frijol de Lima y convertirse en una planta. Para hacerlo, primero remojamos los frijoles durante la noche y luego los envolvimos en una toalla de papel húmeda, colocamos las semillas y la toalla en una bolsa de plástico transparente y la pegamos en una ventana. En casa, puede ayudar a fomentar el aprendizaje de su hijo mediante la repetición de nuestras investigaciones en conjunto. Deje que su hijo tome la iniciativa y que le muestre cómo remojar los frijoles y cómo identificar las diferentes partes de la semilla. Muestre su propio interés y curiosidad con preguntas tales como: "Me pregunto cómo se verá el interior de un frijol- ¿lo sabes? ¿Qué pasa cuando se remojan los frijoles en agua? ¿Cómo podemos ayudar a que la semilla crezca?" Germinen juntos las semillas en una bolsa de plástico transparente. Después de que broten, usted y su hijo pueden plantarlos en una pequeña maceta con tierra y verlos crecer juntos.*

[Note: Be sure to send lima beans home along with this note.]

Assessment—What to Look For

- **Can children describe the lima bean seeds based on observable characteristics?**
  (using senses to gather information, identifying properties, using new or complex vocabulary)
- **Can children predict probable reactions to soaking the seed in water and then compare their predictions to their actual findings?**
  (making predictions, identifying causality, explaining based on evidence, using complex patterns of speech)
- **Can children record (draw or write) observations and contribute to class discussion?**
  (using new or complex vocabulary, documenting and reporting findings, discussing scientific concepts, listening to and understanding speech)

Standards

Head Start Early Learning Outcomes Framework
P-SCI 6. Child analyzes results, draws conclusions, and communicates results. "Analyzes and interprets data and summarizes results of investigation. Draws conclusions, constructs explanations, and verbalizes cause and effect relationships."
Next Generation Science Standards
Science and Engineering Practice: Analyzing and interpreting data. "Use observations (firsthand or from media) to describe patterns and/or relationships in the natural and design world(s) in order to answer scientific questions and solve problems."
***Common Core State Standards* for Mathematics**
1.MD.C.4. Represent and interpret data. "Organize, represent, and interpret data with up to three categories; ask and answer questions about the total number of data points, how many in each category, and how many more or less are in one category than in another."
***Common Core State Standards* for English Language Arts**
W.K.2. Text types and purposes. "Use a combination of drawing, dictating, and writing to compose informative/explanatory texts in which they name what they are writing about and supply some information about the topic."

Terrariums

with Nicole Hawke

Lesson: Planting and observing the growth of plants in a terrarium

Learning Objectives: Children will discuss the needs of plants and use this information to design and construct a terrarium. Children will observe the terrarium over time to identify plant growth and other changes happening inside their terrarium.

Materials: A clear plastic 2-liter bottle for each child; various other containers to compare; writing and drawing materials; (aquarium) gravel; sphagnum or peat moss; activated charcoal (available at your local pet store); potting soil; spoons; clear packing tape; a spray bottle with water; small plants (e.g., begonias, tropical houseplants, ferns) [Note: Your local nursery can direct you to additional plants appropriate for your terrariums and may very well donate plants or cuttings of plants. Be sure to keep leftover plants and grow them at home—use them as your source of plants for next year.]

Safety: Be careful as you cut the plastic bottles that will serve as terrariums; prepare the bottles outside of class time and away from children. Only use pesticide- free and sterilized potting soil and be sure to instruct children not to put their fingers in their mouth or nose while handling potting soil. Make sure any spilled water is wiped up to prevent slipping hazards. Remind children to keep away from electrical outlets when working with water. Make sure children thoroughly wash their hands with soap and water after the activity.

Teacher Content Background: Terrariums are small indoor gardens grown in clear containers. There are a great number of plants that will grow well in your terrariums, which is not surprising, considering that there are hundreds of thousands of different species of plants (estimates typically range between 350,000 and 425,000), with thousands of new species being discovered each year. Biologists classify plants into four broad groups: nonvascular plants (e.g., mosses and liverworts), lower-vascular plants (e.g, ferns and horsetails), gymnosperms (e.g., pines and junipers), and angiosperms (e.g., roses and grasses). Nearly all of these plants have the same four requirements for life: light, air, water, and nutrients (typically provided by soil). Light, air, and water are essential for photosynthesis. In photosynthesis, molecules of water (H_2O) and carbon dioxide (CO_2) are split apart and reorganized—the energy for this comes from light—to produce a unit (CH_2O) used in a molecule of sugar ($C_6H_{12}O_6$), as well as a molecule of oxygen gas (O_2). Similar to how the food we eat is used to support the growth of all parts of our body (muscles, fingernails, etc.), the sugar made through photosynthesis is used to make nearly everything for the plant, as the carbon in sugar, unlike the carbon in carbon dioxide, can be used for all sorts of molecules from cellulose to DNA to chlorophyll. Nutrients provide the other elements required for these molecules and together make plant growth, and therefore life as we know it, possible.

Science terms that may be helpful for teachers to know during this lesson include *investigate*, *observe*, and *terrarium*.

Procedure

Getting Started

Introduction: Take your children to a garden (e.g., a community or botanical garden), or a place where many plants grow at your school, to make observations of plants growing. Let children share with you and with each other the things that they notice about the plants. Then lead children in a discussion about the other places they have seen plants. Some children may have a garden at their homes or may have visited parks where gardens grow; invite children to share these experiences and to share with the class any questions they have about how plants grow.

Prompting Questions: Ask the children what plants need to grow. Encourage children to make observations as you guide them with questions such as "Are there any plants growing on the sidewalk? No? Well, where are the plants growing? Yes, plants grow in the dirt. We call the dirt where plants grow soil; plants need soil. Dig into the soil with your fingers; how does it feel? Is it dry or moist with water? Is the water important? Yes, plants need water."

Back in class, tell children, "Today each of you will make your own terrarium. A terrarium is a place for plants to grow inside our classroom. We have to make sure that our terrariums have everything plants will need." [Note: *Terrarium* is likely a new term for children. Be sure to support their understanding and use of this term throughout the lesson.] Have children explain the things that are needed for their plants in the terrarium to grow. As children share their ideas, display the needs of plants (water, nutrients [soil], air, and sun) in both words and pictures.

Investigating

Comparing: Show children an example (or a picture) of a terrarium—to encourage children's creativity, make sure your sample terrarium is made of a container different from the possible containers you share with children. Explain to children, "I have plants and all the materials you'll need for your terrarium, but first we have to decide which type of container to use." Present children with several different options such as empty ½ gallon milk cartons, shampoo bottles, paper bags, clear, plastic bottles, cereal boxes, and coffee cans. Make sure that you include containers that are opaque, ones that are translucent, and one container (plastic 2-liter bottle) that is transparent. Prompt children to consider which container would be best for their plants and let them discuss. If needed, prompt further debate by posing questions such as, "Plants need water; which containers can hold water? Plants need light; which containers will give the plants the most light?" Once children have selected a clear plastic bottle as the best container for their terrariums, ask them what you'll need to put inside the bottle to make a terrarium. Help them to think once again about what plants need. The plants in your terrariums will need air, light, water, and soil. [Note: If some children choose a different container, that's fine; build a terrarium in that container and compare it and its plants' growth to other terrariums later in the lesson. In fact, you may choose to do this as you make your terrarium, so that children can later compare and describe how their design is better than yours.]

Now that children have considered and discussed what plants need to grow, it is time to put this knowledge to work as they construct their own terrarium. (When children apply what they've learned, it helps to solidify their new understanding and gives purpose and meaning to their learning.) For their terrarium, each child will need a clear plastic bottle with the top third of the bottle removed. [Note: Prepare the bottles beforehand, being very careful as you cut each bottle. Use stickers or tape to mark the top and bottom of the bottle so you can match the correct halves later.] The construction of the terrarium can be set up as centers that the children rotate through, adding a different material at each, or construction can be completed at one location. Either way, be sure to have the children do as much of the construction as possible. Below are the steps to follow in order to construct each terrarium.

1. Put a thin layer of gravel at the bottom of a clean container. Spaces between pieces of gravel will provide drainage and air pockets for roots.
2. Add a thin layer of sphagnum or peat moss. This serves to keep the soil from falling to the bottom of the terrarium.

Making a terrarium

3. Next, add a thin layer of activated charcoal. This layer will help to absorb odors and keep your terrarium fresh.
4. Add 2 to 3 inches of potting soil.
5. Have children use a spoon to make a hole in the soil for each plant they will place in their terrarium.
6. Place each plant in a hole and gently press the soil in around the plant. Add additional soil around the plants as needed. [Note: If you prefer, children can plant seeds rather than plants. Add an additional inch of potting soil in Step 4 and follow the planting instructions on the back of the seed pack.]
7. If you'd like, children can add a few rocks, plastic figures, or other decorations to the surface of the soil.
8. Use a spray bottle to add water to the soil, making it moist, but be cautious to not overwater. The entire surface of the soil should be wet, but without puddles of standing water.
9. Seal your terrarium to help keep moisture in the terrarium, but don't seal it so perfectly that it is airtight—you'll want some airflow. You may have to complete this step for children. Use clear packing tape to reconnect the bottle. Be sure that the cap is on the bottle. Sealed correctly, terrariums will rarely need additional watering. [Note: Although it is unnecessary, you may choose to poke one or two small holes in the side of each terrarium using a small tack or pin. The holes will serve as a visible source of air for the plants inside, helping children to see how all the needs of plants are met.]
10. Make sure children wash their hands thoroughly upon completion of their terrarium.

Store the terrariums in a place with indirect sunlight—direct sunlight can overheat the terrariums and harm the plants inside.

Measuring and Documenting: Once the terrariums are complete, collaborate with children to record what the plants in their terrarium look like initially. Help them use a ruler to measure the height of their plants (or, depending on the plants you use, children can count the number of leaves). Photograph each child's terrarium. Print this picture with space on the page for the child to write his or her name, the height of the plant (or the number of leaves), and the date that the photograph was taken (see Figures 3.4 and 3.5). Keep these data sheets displayed on a poster or in a notebook near the terrariums. Allow children opportunities to look at the terrariums and review the data sheets.

Making Sense

Describing Findings: After a few weeks, have children share the progress of their terrariums. [Note: The timing of this stage of the lesson will depend on the plants you select for your terrariums. Before this part of the lesson, be sure that there is noticeable plant growth for children to observe, measure, and report.] Have children verbally share with the class the things they notice are changing and other things that are staying the same. Help each child to measure the height of his or her plant or to count the leaves. Take a photograph and print a new data sheet for children to complete. Prompt children to compare their initial and new measurements to determine which is larger. Children will be excited to see that their plant is growing. Remind children that this growth is only possible if plants have the four things that plants need to grow; ask children if they can remember the things that plants need and to identify how the plants in their terrarium are meeting each need. Encourage children to compare their terrariums with each other, noting what things are happening in both terrariums and comparing the heights of their plants (see *Math Connections*, p. 106).

What's Next?

Extension Activity

A day or two after constructing your terrariums, have children make observations of their terrariums. Although plants in the children's terrariums will be unchanged, there will be one very noticeable change that occurs. Ask children to look for the water in their terrarium. Remind them that when they made the terrariums they sprayed water on the surface of the soil. Ask, "Where can you see water now?" Children may be surprised to find water droplets along the upper surfaces of the bottles. Encourage children to watch the drops of water to see where they go. The inadvertent jostling, as children hold and look at the terrariums, will cause several of the drops to slide down the plastic to the soil below. Ask children why it might be important for the plants that the waterdrops fall down to the soil and connect this idea to plants growing outside, asking, "Do waterdrops fall down to the soil for plants outside too? How do plants growing outside get their water?" [Note: For terrariums with a lot of condensation, open the bottle cap and let the terrariums sit overnight so that some of the water evaporates out of the bottle. Replace the cap the next morning. For terrariums with little to no condensation, take the cap off the bottle and spray some additional water into the terrarium and replace the cap. Reassess the moisture content of the terrariums the next day. It's a good idea to assess the condensation in the terrariums regularly for the first few weeks and monthly after that, adding water as necessary.]

Figure 3.4

Terrarium Data Sheet 1

Insert picture of plant here	Name
	Date
How tall is my plant? ______ inches	

Figure 3.5

Terrarium Data Sheet 2

Insert picture of plant here	Name
	Date
How many leaves does my plant have? ______ leaves	

Full-size versions of these figures are available at www.nsta.org/startlifesci.

Integration to Other Content Areas

Reading Connections

The terrarium investigation takes place over a long duration of time, providing ample opportunity to introduce quality children's literature related to plants and gardens. There are a great many enjoyable books that share stories about gardens and the joy that planting and tending a garden can bring. *The Curious Garden* by Peter Brown (2009), *The Gardener* by Sarah Stewart (2007), *The Night Gardener* by Terry and Eric Fan (2016), and *Wanda's Roses* by Pat Brisson (2000) include characters who love to garden and might serve as role models to the young gardeners in your class, inspiring in children a greater appreciation for and an interest in gardens and the beauty they provide. And many other books use narrative to describe the process for planting and caring for a garden, including *Zinnia's Flower Garden* by Monica Wellington (2007) and *Eddie's Garden and How to Make Things Grow* by Sarah Garland (2009). (See *Nature Bracelets*, p. 144, for more book ideas.)

Writing Connections

The data sheets used in the lesson ask for children to record their names, the date, and the plant height or number of leaves. You can add to this a place where children can write a sentence describing their terrarium. Encourage children to read their sentences to you. Students can also draw a picture of their terrarium and use words to label the different components (e.g., *plant, soil, water, sun*). This labeling can help children to begin to understand the meaning of written words. Also, if children included small plastic figures in their terrarium, they could make up a story about the figures and their "life" in the terrarium. Children can dictate this story to you or write it themselves in pictures (or words) and then "read" their story to you. This type of storytelling helps to develop oral fluency, vocabulary, and expression and can be an important bridge to writing.

Math Connections

Your terrarium plants and data sheets can inspire all sorts of plant counting and measuring. Take

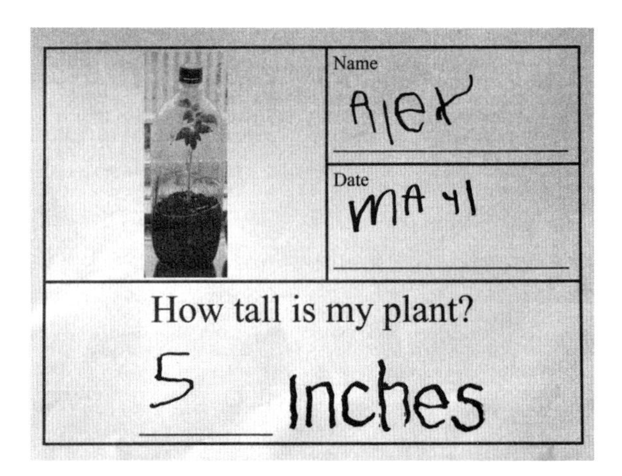

Sample terrarium data sheet

children outside and encourage them to count the number of flowers on a bush, the number of petals on a flower, the number of dandelions in the grass, or the number of trees in the school yard. Using both standard and nonstandard units, children can measure various plants and plant features. Prompt children to measure and compare plants, asking, "Who can find the biggest leaf? Which tree is the biggest around? Can you find a dandelion with a stem longer than this one?"

Other Connections

Child's Life Connections

Each day, children have the opportunity to observe a great diversity of plants. Children can count, and even photograph, all the different types of plants on the school grounds, in their homes, or at a nearby park and keep a plant journal where they can record the different plants they see. Help children to appreciate the diversity of plant life around them by sending them on a plant scavenger hunt. Describe, name, and show pictures of a particular plant that is found on the school grounds. Ask children to explore to find the plant. Challenge them to find a new plant each day. A variation of this scavenger hunt is one focused on plant colors. Give children paint swatches (available for free at paint or home improvement stores) and encourage children to find plants that match the color. Challenge children to find several different matches. [Note: Be sure to select color swatches that correspond to the plants on your school grounds.]

Center Connections

In your **art center**, provide children with different magazines with pictures of plants. Have children create a picture of a garden. Have the children cut out and glue plants they like and have them verbalize why they chose certain plants to be a part of their garden. At the **sensory table**, add examples of the plants children planted in their terrariums. Provide children with magnifiers and encourage the children to look closely at the leaves and other visible parts of the plants. Help children to look closely by asking question such as, "Do some of the plants have leaves that are hairy? Are any of the leaves smooth?" For **dramatic play**, children can create fairy gardens. Provide boxes and various small plants, rocks, and sticks for children to arrange to create a miniature garden. Also provide small figurines for children to place in their garden. Children can then imagine, act out, and tell each other about the adventures the characters experience in their garden.

Family Activities

In school, your child has been learning about different types of plants. You can further your child's interest in plants by talking with your child about the different places that have plants and going to observe plants together. Visit a local park or garden. Together with your child, look for ways that plants are similar and ways they are different. Engage your child in excited conversations about your discoveries. For example, "This tree has long, narrow leaves; does that tree have long, narrow leaves? How about that other one over there?" and "What color is that flower? Are there other plants that have flowers? What colors are those flowers?" Together, identify your favorite plants and explain your choices. On the way home, stop at the library and check out books about plants to read together. As you read, stop and explore the pictures together. Ask your child questions such as "How many plants are on this page? What color are these flowers? Which of these is your favorite?"

Actividades Familiares

En la escuela, su hijo ha estado aprendiendo sobre los distintos tipos de plantas. Puede fomentar el interés del niño por las plantas hablando con él sobre los diferentes lugares diferentes que tienen plantas y salgan a observar las plantas juntos. Visite un parque local o jardín. Junto con su hijo, busque formas en que las plantas son similares y diferentes. Participe con su hijo en conversaciones animadas sobre sus descubrimientos. Por ejemplo: "Este árbol tiene hojas largas y angostasestrechas; ¿ese árbol tiene hojas largas y angostasestrechas? ¿Qué hay de ese otro de allá?" Y "¿De qué color es esa flor? ¿Hay otras plantas que tienen flores? ¿Qué colores son esas flores?" Juntos, identifiquen sus plantas favoritas y expliquen sus elecciones. De camino a casa, deténganse en la biblioteca y pidan libros sobre plantas para leer juntos. Al leer, hagan una pausa y exploren las imágenes juntos. Haga preguntas a su hijo tales como: "¿Cuántas plantas hay en esta página? ¿De qué color son las flores? ¿Cuál de ellas es tu favorita?"

Assessment—What to Look For

- **Can children describe plants based on their observations?**
  (using senses to gather information, identifying properties, using complex patterns of speech)
- **Can children measure, record, and describe the growth of plants in their terrarium?**
  (comparing properties, measuring, documenting and representing findings, identifying change)
- **Can children provide verbal comparisons of the plants month by month?**
  (comparing properties, using complex patterns of speech, discussing scientific concepts)

Standards

Head Start Early Learning Outcomes Framework
P-SCI 6. Child analyzes results, draws conclusions, and communicates results. "Analyzes and interprets data and summarizes results of investigation. Draws conclusions, constructs explanations, and verbalizes cause and effect relationships."
Next Generation Science Standards
Science and Engineering Practice: Analyzing and interpreting data. "Record information (observations, thoughts, and ideas). Use and share pictures, drawings, and/or writings of observations."
Common Core State Standards for Mathematics
1.MD.A.2. Measure lengths indirectly and by iterating length units. "Express the length of an object as a whole number of length units, by laying multiple copies of a shorter object (the length unit) end to end; understand that the length measurement of an object is the number of same-size length units that span it with no gaps or overlaps."
Common Core State Standards for English Language Arts
W.K.2. Text types and purposes. "*Use a combination of drawing, dictating, and writing to compose informative/explanatory texts in which they name what they are writing about and supply some information about the topic.*"

Smelling Plants

with Emily Kraemer

Lesson: Exploring the scents of different plants

Learning Objectives: Children will understand that plants have distinct odors by observing different plant scents and characterizing these different types of scents (sweet smells, fruity smells, etc.).

Materials: Five sets of "smelly cans" (opaque containers each with one plant scent inside—film canisters work well), five sets of picture cards that represent the items inside the canisters (include the name of the item on each picture card), writing and drawing materials

[Note: In making your "smelly cans," use scents such as lemon, cinnamon, vanilla, garlic, mustard, pine, lavender, apple, dill, ginger, mint, or banana. Crush a small piece of each plant part and place it in a canister. If the scent is not strong enough, you can place a cotton ball that has been wetted with oils or extracts of the scents inside each canister. Poke holes in the lids of the canisters (or cover the tops of the canisters with aluminum foil and poke holes in the foil) so that children can smell, but not see, inside. It will be helpful if the children have prior experience with the items being placed inside the canisters. If your children are from a culture where certain spices are used (such as allspice, curry powder, cumin, etc.) it might be helpful to include these spices in the canisters. It will also be helpful if children have prior experience categorizing items. You may want to have children practice sorting objects into groups of their own choosing beforehand; see, for example *Sorting Seeds*, page 75.]

Safety: Allergies to several different plants and plant parts are possible. Do not use nut-based products like peanut butter and avoid spices that are too pungent or "hot," such as cayenne pepper. Be sure to find out if any of your children have food allergies before using food in your classroom and remind yourself of the symptoms of allergic reactions in children. Also, remind children not to taste or eat the contents of the canisters.

Teacher Content Background: Plants can't move. Certainly they can bend, twist, sprawl, climb, and grow, but (with the exception of a few aquatic plants) they are literally rooted in place. Being stationary has a number of challenges; for the most part, these challenges can be thought of as difficulty with getting away from something bad and difficulty with getting to something good. The pungent smells of some plants are often produced to protect the plant by discouraging herbivores. People sometimes use small amounts of these aromatic plants as herbs and spices. (To understand how these flavorful chemicals that we intentionally add to our food could deter herbivores, imagine a mouthful of cinnamon.) The flavor of sage, rosemary, basil, mint, marjoram, and many others comes from oils secreted or stored on the surface of the plant that encourage aphids, caterpillars, and grasshoppers to steer clear. However, sometimes plants want to attract insects (and birds and even mammals) as they can be helpful when it comes to seed dispersal (moving seeds away from the plant that produced them) and pollination (transferring pollen from one flower to another). Many fruits have odors that encourage animals to eat them and in the process scatter seeds. Most of the fruits that we eat, from tomatoes to tangerines, serve this purpose. Similarly, the sweet scents produced by flowers such as jasmine, violets, and roses are designed to attract pollinators to the flower. For centuries, people have used the chemicals that produce these *attractive* smells as perfumes. Of course, the aroma of the carrion flower and corpse flower is not exactly pleasant—these plants attract flies to serve as pollinators by smelling like dead, decaying animals.

Science terms that may be helpful for teachers to know during this lesson include *sorting*, *observe*, and *scent*.

Procedure

Getting Started

Introduction: To start this lesson, share some smells with children. Before class, cover the labels of two scented air fresheners. Spray the first one and model for the children your observations of the scent, focusing on descriptions more than identification. For example, "Oh, this smells fruity. This smells sweet." Then spray the second air freshener and guide children to make observations with you. Encourage children to use full sentences (i.e., "This smells ____") and record the different adjectives and descriptions they use. The goal here is for you to model and for children to practice describing scents that the entire class can smell. [Note: If air fresheners are discouraged at your school, you can use extracts or essential oils. Place a few drops on an index card and gently wave the card in front of children.]

Curiosity: To further spark children's interest, ask them about some of their favorite smells, again encouraging children to describe, not just identify, the scent. After all the children have had a turn to share, ask them to describe some of their least favorite or "stinkiest" scents. Throughout the conversation, record the different adjectives and descriptions children use.

Investigating

Observing: Distribute sets of "smelly cans" to small groups of children. Allow children to freely explore the canisters, encouraging them to smell the different canisters, but ask that they do not open the canisters. This open exploration is important before moving to more directed explorations, as it provides opportunities for child thinking and creativity and it provides experiences that children can draw from as they engage in subsequent, more structured learning.

Sorting: Once children have had an opportunity to explore on their own, ask children to place the canisters into groups that they think go together based on their scents. Explain that this process of putting similar things into groups based on their characteristics is called *sorting* and remind children of previous activities that required them to use the skill of sorting. (See *Sorting Seeds*, p. 75, for more information on sorting.) Let children know that there are several different ways to sort and encourage them to find multiple ways to sort the canisters (e.g., grouping together all the cans that smell fruity or stinky or spicy).

Making Sense

Describing Findings: Next, tell children you are excited to find out what kinds of groups they sorted their smelly cans into and ask them to explain their groups. Record the names or descriptions of the groups (e.g., spicy, sweet, yummy) that children offer. To encourage an expanded list of descriptors, ask children if they can sort their smelly cans into new groups. Again, ask children to explain their groups and record their descriptions.

Application: Provide children with the picture cards of different plants and ask them to match the smelly can to the picture. After they've attempted to pair pictures with cans, allow each child to open the cans to see if he or she found any matches. [Note: The goal here is for children to use the new descriptors of scents and to

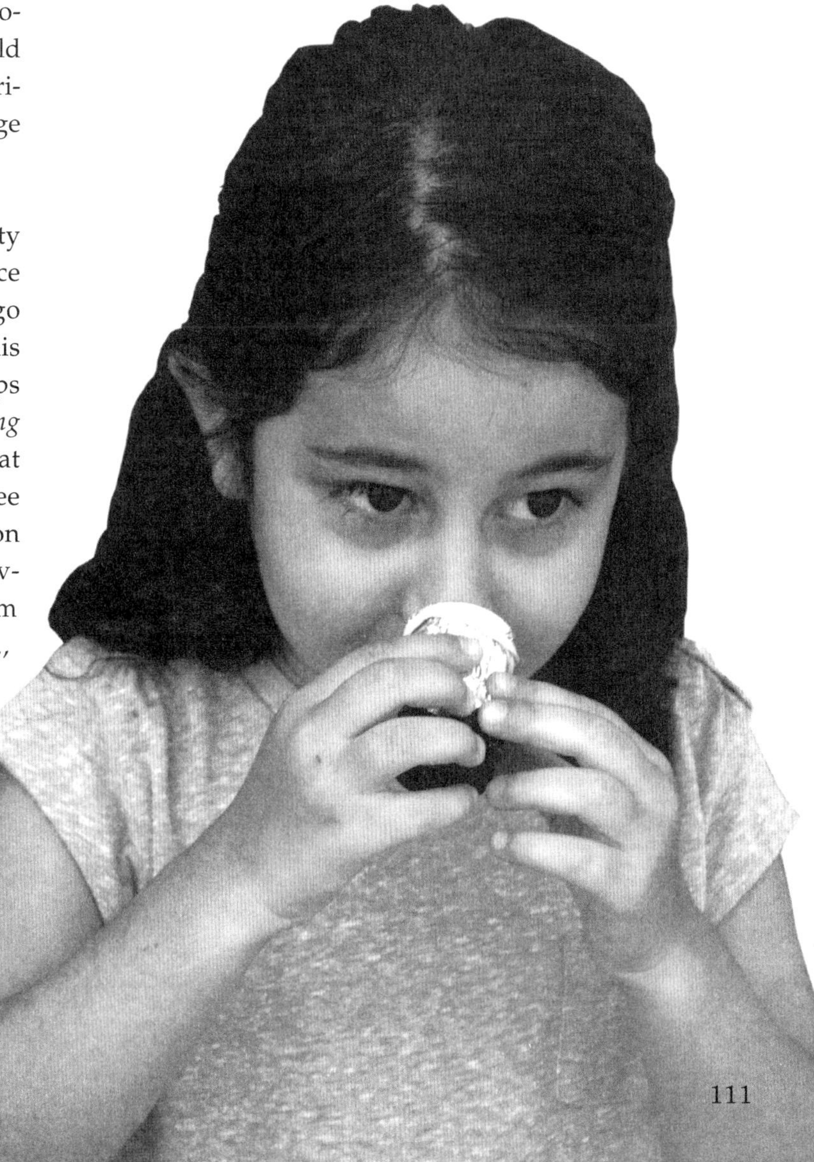

understand that plants make different scents; if children don't correctly match pictures to scents, that's fine. The experience is much more important than learning plant names.] Encourage children to describe the scents and the plants shown on the picture cards. As a group, use some of the descriptors invented by the children, asking, "Which of these plants has a scent that is ________?" Record children's responses or attach pictures of plants to a chart with a column for each type of scent. Ask the children what other things they think would fit into each group; add their ideas to the chart. After children have shared their ideas, show children a few additional items that match the scent headings of your chart (e.g., an old shoe, a cupcake, etc.). Have children decide where these new objects should be placed on the chart.

What's Next?

Extension Activity

You can encourage children to make further explorations with their sense of smell by engaging them in a smelly cans scent-matching activity. In this activity give children two identical sets of smelly cans. Be sure that the cans are not labeled. Ask children to smell each and to find the two cans that smell alike. As an alternative, you can give children one set of smelly cans labeled with a picture of the contents and one unlabeled mystery can. Children can then try to find the scent that matches the mystery can. Repeat with a new mystery can until all smelly cans have a match. Throughout this activity, use questioning to encourage children to describe the scents, "How are these two scents different? Why do you think the mystery can is pineapple? How would you describe this scent?" These open-ended questions will help to develop children's critical thinking and oral language skills.

Integration to Other Content Areas

Reading Connections

Within the lesson and extension activity, children use descriptive oral vocabulary that supports the development of reading skills. During reading time, share scented books such as *Little Bunny Follows His Nose* by Katherine Howard (2004) and *Scratch and Sniff: Garden* and *Scratch and Sniff: Food* (1999; 1999) by DK Publishing. These and similar books can help reinforce the scents discovered in the lesson, as well as introduce children to additional scents. Be sure to share books that depict some of the different plants that you used in the *Investigating* section of the lesson. For example, if you used apple as one of the scents in the lesson, *How Do Apples Grow?* by Betsy Maestro (2000) and *Apples Grow on a Tree* by Mari Schuh (2011) may be helpful, as these and similar books can help children to gain a fuller appreciation of the plants they investigated. Books about the five senses, such as *Hello Ocean* by Pam Muñoz Ryan (2001); *My Five Senses* by Aliki (2015); and *Look, Listen, Taste, Touch, and Smell: Learning About Your Five Senses* by Pamela Hill Nettleton (2004), can help children to understand how smell is just one way that we can learn more about the world around us.

Writing Connections

Make a class book titled "Our Favorite (Plant) Scents." For each child, divide a piece of construction paper into three sections; each child's paper will become the pages of your class book. In one section of the page, children can paste a picture of the plant that makes their favorite scent. In another section, tape a small, sealable plastic bag to the page. Children can place some of the leaves, fruits, or flowers that produce their favorite scent in their bag. For the final section, ask children to tell you a story about the scent they chose and write down their words. Compile children's pages into your class book and leave the book out for children to read and explore. [Note: Depending on the scents children select, you may have to replace the bags and their contents every few days.]

Math Connections

Create a bar graph showing your whole class's favorite scent. Provide multiple sets of picture cards that represent five different scents (e.g., banana, cinnamon, mint, pine, rose flowers) and the smelly cans for each scent. Have each child decide which can they think smells the best and select a picture card that represents their favorite scent. Children can color their picture cards and then place their cards onto a class graph. During group time, when the graph is complete, have the children count the votes for each scent and write the number in each column. Help children to understand concepts of equality by encouraging them to identify the scents with the same, most, and least amount of votes and to compare the votes for two scents to decide which received more votes and which received fewer votes.

Other Connections

Child's Life Connection

There are so many different scents that children encounter throughout each day. Challenge children to pay closer attention to the scents in their life. Children can keep a scent diary, drawing or taking pictures of the different scented things they discover. Each individual child can keep his or her own scent diary, or you may choose to have a class diary that children take turns making entries in. Or have a scent scavenger hunt that requires children to find things outside of school that have a particular scent—use the descriptors from the chart you made in the *Making Sense* portion of the lesson. When children return to class, have them share entries from their scent diary or things that they found during their scavenger hunt.

Center Connections

In the **art center**, set out flowers in a vase. Provide children with paper, brushes, and scented

watercolors and let them create their own scented painting of the flowers. [Note: There are many different recipes for creating scented watercolors; the most simple is just adding food coloring to water along with a drop or two of scented extract. Extracts such as vanilla, lemon, and strawberry are available in the baking aisle of most grocery stores.] For the **sensory table**, add food coloring and different scents to your favorite homemade playdough recipe. For scented playdough, you can add different extracts or add spices and fruits from your pantry (ground cinnamon, banana, lemon zest, etc.); or you can use different flavors of powdered drink mix or gelatin. Let children play with this scented playdough and encourage them to describe the way it looks, feels, and smells. (Remind them not to eat the playdough, no matter how delicious it smells.) Children can be chemists in the **dramatic play** center. Provide children with lab coats, goggles, and gloves and set out different scent cards (index cards with a drop or two of scented extract along with a picture and the name of the scent) and cups with perforated lids to hold the cards. Children can mix scents together, placing different combinations of the scented cards in the cups, and then shake and smell to discover new scents.

Family Activities

Your child has been using her or his sense of smell to explore different scents. Ask your child to share the scents that we've explored at school and ask if she or he has a favorite scent. Share your favorite scent (think about smells that are unique to your child's home or culture—spices, perfume, incense, flowers) and then explore together, asking other family members or friends what their favorite scents are. If someone can't think of a smell, your child can describe some of the other scents that people have shared. Your child can keep a "list" of pictures that show each person's favorite scent. Throughout the week, you can use your sense of smell to explore, drawing your child's attention to different scents and describing together with your child how each smells. Be sure to only use natural scents, such as foods and plants, and avoid all chemical odors, such as household cleaners.

Actividades Familiares

Su hijo ha estado utilizando su sentido del olfato para explorar diferentes aromas. Pídale a su hijo que comparta con usted los olores que nos hemos explorado en la escuela y pregúntele si tiene un olor preferido. Comparta su aroma favorito (piense en los olores que son únicos en su hogar o cultura—especias, perfumes, incienso, flores) y luego, exploren juntos, preguntando a otros miembros de la familia o amigos cuáles son sus aromas favoritos. Si alguien no puede pensar en un olor, su hijo puede describir algunos de los otros aromas que otras personas han compartido. Su hijo puede llevar una "lista" de fotos las imágenes que muestren el aroma favorito de cada persona. A lo largo de la semana, puede usar su sentido del olfato para explorar, llamando la atención de su hijo o para descubrir aromas diferentes y describiendo juntos como huele cada uno. Asegúrese de utilizar solamente aromas naturales, como de alimentos y plantas y evite todos los olores químicos, tales como productos de limpieza adores para el hogar.

Assessment—What to Look For

- ***Can children describe scents based on their observations?***
  (using senses to gather information, identifying properties, using complex patterns of speech)
- ***Can children sort items based on observable characteristics?***
  (identifying properties, comparing properties, classifying)
- ***Can children explain their reasoning?***
  (constructing explanations, explaining based on evidence, using complex patterns of speech)

Standards

Head Start Early Learning Outcomes Framework
P-SCI 3. Child compares and categorizes observable phenomena. "Categorizes by sorting observable phenomena into groups based on attributes such as appearance, weight, function, ability, texture, odor, and sound."
Next Generation Science Standards
Science and Engineering Practice: Planning and carrying out investigations. "Make observations (firsthand or from media) and/or measurements to collect data that can be used to make comparisons."
Common Core State Standards for Mathematics
K.MD.B.3. Classify objects and count the number of objects in each category. "Classify objects into given categories; count the numbers of objects in each category and sort the categories by count."
Common Core State Standards for English Language Arts
L.K.5.A. Vocabulary acquisition and use. "Sort common objects into categories (e.g., shapes, foods) to gain a sense of the concepts the categories represent."

Pumpkin Outsides

with Susan Gomez Zwiep

Lesson: Observing, measuring, and comparing whole pumpkins

Learning Objectives: Children will observe, describe, and compare pumpkins and generate a list of ways individual pumpkins are similar and different.

Materials: Multiple pumpkins of various shapes, sizes, and colors; one strip of paper for each pumpkin (each strip should be about 2 inches wide and at least as long as the circumference of the widest pumpkin); writing and drawing materials

Safety: Although rare, food allergies to pumpkin, including pumpkin seeds, are possible. Find out if any of your children have allergies before using the materials in this lesson and remind yourself of the symptoms of allergic reactions in children. Remind children not to eat any food being used in the activity and not to put their fingers in their mouth or nose while handling plant material. Make sure children thoroughly wash their hands with soap and water after the activity.

Teacher Content Background: Pumpkins, like zucchini and all kinds of squash, cucumbers (and therefore pickles), and all melons, belong to the plant family *Cucurbitaceae* (also known, as the *Cucurbits*). Pumpkins are distinguished from their *Cucurbit* relatives by their large size and often bright colors. Although our "typical" pumpkins are medium to yellowish orange in color, mostly round, and smooth with shallow ridges, there are dozens of different varieties of pumpkin with interesting names such as Jumpin' Jack, Cinderella, Little Boo, Sugar Pie, and Red Warty. Across these varieties, pumpkins can come in a whole range of colors (including green, yellow, white, blue, and red), textures (including smooth, ribbed, and bumpy), and sizes. An average-sized pumpkin weighs about 13 pounds (6 kilograms), but the pumpkins grown for "largest pumpkin" competitions regularly exceed 1,000 pounds (450 kilograms)—the current world record pumpkin weighed more than 2,600 pounds (1,190 kilograms)! (See *Pumpkins Insides*, p. 122, for more information about pumpkins.)

Science terms that may be helpful for teachers to know during this lesson include *measure*, *observe*, *fruit*, and *pumpkin*.

Procedure

[Note: Although pumpkins are used here, many of these activities can be completed with any fruit. The advantage of pumpkins is their large size, making them easier for children to manipulate and observe.]

Getting Started

Prior Knowledge: Set out a few different-sized pumpkins in front of the group to introduce children to the lesson. Ask children if they know what these objects are and where they come from. Have children share their ideas and, if needed, tell the children that the objects are all pumpkins and that pumpkins come from pumpkin plants. Use open-ended questions to extend the conversation by asking, "Have you ever used a pumpkin? How? Where have you seen pumpkins before? Have you eaten foods with pumpkin? What kinds of food? Have you seen a pumpkin growing? Has anyone ever grown a pumpkin? What questions do you have about pumpkins?" This type of opening conversation about a topic helps children to become mentally ready for and interested in further exploration about the topic.

Investigating

Observing: Give each pair of children a pumpkin to explore. Invite the children to use multiple senses as they observe the pumpkins. [Note: Be sure to provide a variety of pumpkins with obvious differences. See *Teacher Content Background* section for some of the characteristics of different varieties of pumpkins.] As children observe, encourage them to describe their pumpkin by asking questions such as, "What does it look like? What does the texture of your pumpkin feel like? What does your pumpkin sound like if you tap it with your hand?" Once children have had time to observe their pumpkin, give them each a large (12" × 18") piece of paper that has been divided in half. Have children record their observations by drawing their pumpkin on the left side of the paper.

Comparing: Now that children have observed and described their pumpkin, ask children to carefully observe a different pumpkin and to consider how the two pumpkins are different from each other. Have each pair of children partner with another pair to compare their two pumpkins. [Note: Be sure groups have two very different pumpkins to compare.] Encourage the children to look for and to discuss ways that the pumpkins are different from each other. When needed prompt children with questions such as "Do they have the same color? How do they feel?" Children can make a drawing to record observations of the new pumpkin on the right side of their paper and as they do so, encourage them to describe how the second pumpkin is different from the first.

Making Sense

Describing Findings: Bring the children together and pass the pumpkins around, giving time for children to observe each pumpkin, and then place the pumpkins in front of the children. Draw an outline of a pumpkin on chart paper and beneath the outline make two columns: one labeled *same* and one labeled *different*. Help children to read these labels by emphasizing first-letter recognition and letter sounds. Then ask children to share what they observed about their pumpkins. As a child describes a characteristic, have the other children look at the pumpkins and decide if all the pumpkins fit that description. If it is a characteristic that all the pumpkins have, add it to the *same* column, but if not all the pumpkins have it, add it to the *different* column. For example, if a child says his or her pumpkin is orange, ask children if all the pumpkins are and to describe the other colors they see, then add *orange* to the *different* column. Continue observing and comparing the different characteristics of pumpkins, adding words with picture support to your *same/different* chart. When your chart is complete, help children to understand that pumpkins have certain characteristics in common (the *same* column in your chart), but can also have individual differences (the *different* column).

Measuring a pumpkin

Application: After exploring and discussing pumpkins, children can apply their understanding that things can have individual differences but still be the same to other things in the classroom. Let children explore, backpacks, shoes, crayons, and so on to find, for example, that individual backpacks have differences (e.g., size, color), but that they all have things in common (e.g., have straps, carry books). Provide children with opportunities to explore living things as well, to find that—just like pumpkins—trees, teachers, dogs, classmates, etc. have characteristics that are the same and characteristics that are different.

What's Next?

Extension Activity

Pass out a strip of paper to each child and challenge children to figure out how to use the strip of paper to measure the distance around a pumpkin. If needed, show the children how to wrap the strip around the widest area of a pumpkin and how to draw a line where the ends of the paper overlap. Have children cut across the line so they have a paper strip that is as long as the pumpkin is big around. Have children write their names on their pumpkin strips so that your class can tell which strip fits which pumpkin. See if any pumpkin strips fit around any of the children's waists or chests or heads. Ask the children to compare the strips from each pumpkin to find out which ones were similar in size and which ones were different. Hang up the pumpkin strips to show the different sizes. Invite children to help you order the strips from the largest to the smallest pumpkin, asking, "Which pumpkin strip belongs to the biggest pumpkin? The smallest pumpkin?" (See the *Math Connection* section for ways to continue to use your pumpkin strips.)

Integration to Other Content Areas

Reading Connections

It seems like there are more books about pumpkins than about any other fruit! There are many great nonfiction books that can help children to learn more about pumpkins, such as *Pumpkins* by Jacqueline Farmer (2004) and *The Pumpkin Book* by Gail Gibbons (1999). And there are many stories about pumpkins, as well. There are stories about growing pumpkins: *It's Pumpkin Time* by Zoe Hall (1999), *Pumpkin Circle: The Story of a Garden* by George Levenson (2002), and *The Very Best Pumpkin* by Mark Kimball Moulton (2010). There are stories about chasing pumpkins: *The Runaway Pumpkin* by Kevin Lewis (2005). There are also stories about decorating pumpkins: *It's Pumpkin Day, Mouse!* by Laura Numeroff (2012). (See *Pumpkins Insides*, p. 122, for more pumpkin-related books.) Perhaps the most interesting books about pumpkins focus on what happens to jack-o'-lanterns after Halloween. Check out the books *Pumpkin Jack* by Will Hubbell (2000) and the more detailed *Rotten Pumpkin: A Rotten Tale in 15 Voices* by David M. Schwartz (2013) as well as the *Writing Connections* below to learn more about pumpkin decomposition.

Writing Connections

One of the best things to do with a pumpkin is to carve it and then watch it slowly decompose. You can share this amazing experience with your children and together make a book documenting a pumpkin's decomposition. Show a time-lapse video of this captivating (i.e., gross) transformation with children and set up your own pumpkin to watch it rot in real time. Carve several holes in a small pumpkin and let it sit out in your room for the first day or two, then transfer it to a clear, sealed, disposable container. Each day, have a child take a picture of the pumpkin. Print this picture with space on the bottom of the page for the child to write the day of the investigation (e.g., "Day 2") and for you to write a brief description that a child dictates to you. After each child has written a page (or once your pumpkin is just too gross to continue), assemble the pages together in a book. Let your children help you arrange the pages in the appropriate order and then ask them for ideas for a book title. Title the book and place it in the reading center. [Note: Do not open the container as many of the fungi and other decomposers at work on your pumpkin release airborn spores. At the end of this activity, put the entire closed container into the trash.]

Math Connections

Children can take their pumpkin strips from the extension activity and compare them to other objects in the room. "My pumpkin strip is longer than my book." "My pumpkin strip is shorter than my friend." Children can use this same indirect measure to compare short distances such as from their desk to the next desk or to some other location in the room. Direct children to place one end of their pumpkin strip at the starting point and have another child mark the end of the strip. The first child then moves the strip to that mark to continue measuring. They can then describe distances in relative terms, such as "From my desk to her desk is longer than from my desk to the teacher's desk," or by the number of pumpkin strips: "From the block area to the door is six pumpkin strips long."

Other Connections

Child's Life Connection

Pumpkins are related to a great number of other edible plants—so many that your children are all but guaranteed to be familiar with at least one (see *Teacher Content Background* section). Set out a pumpkin and a number of pumpkin "relatives." Invite children to explore the fruits, searching for commonalities as well as differences. As they observe, encourage them to use their sense of touch as well as sight. Have children sort the fruits into groups that make sense to them. Maybe they'll group the pumpkin, zucchini, and acorn squash together—after all, scientists do, classifying these three, along with yellow, crook-necked, gourd, and a few other squash, as all different varieties of the same exact species! However they sort the fruits, applaud their efforts and have children explain to you the thinking behind their sorting.

Center Connections

Use pumpkins in a variety of projects in the **art center**. Pumpkins can be the canvas that your children paint on. Invite children to paint "pumpkin faces" using the stem of the pumpkin as the nose. Or ask children to use pumpkins as their brushes. Children can dip their pumpkins in paint and blot this paint on paper. Provide several different pumpkins to allow children to make prints of different sizes and shapes. Or children can use pumpkin pulp as the medium for their art. Children can paint pictures in paste and then adhere chucks and strands of pumpkin pulp to their pictures. Use pumpkin seeds at the **sensory table**. Provide small tubs of pumpkin seeds prepared in different ways. Use slimy seeds fresh out of a pumpkin, clean pumpkin seeds, and roasted pumpkin seeds. Encourage children to experience the different textures of the seeds. At the **dramatic play** center children can pretend to be pumpkin farmers, planting, growing, harvesting, and selling pumpkins. Put out farmer clothes, toy garden tools, and lots of pumpkins. As the children play, ask them what other materials they need to pretend to be pumpkin farmers.

Family Activities

At school, your child has been studying pumpkins. You can ask your child about the pumpkin activities he or she has been doing, and you can encourage your child to learn even more about pumpkins by taking a family trip to a local farm, pumpkin patch, or market to observe and discuss pumpkins. While you are there, compare different pumpkins. Encourage your child to tell you what two pumpkins have in common and to point out any differences he or she notices. Use your child's descriptions to inspire you to search together for the largest and the smallest pumpkin. You can also search for the pumpkin that is the bumpiest, flattest, tallest, roundest, and so on. Be enthusiastic as you follow your child's lead and discover pumpkins together.

Actividades Familiares

En la escuela, su hijo ha estado estudiando las calabazas. Puede preguntarle a su hijo sobre las actividades de la calabaza que ha estado realizando en torno a la calabaza yhaciendo y usted puede animar a fomentar que su hijo a que aprenda aún más sobre ellas llevándolo a en un paseo familiar a una granja local, a un huerto de calabazas o al mercado para observar analizar las calabazas. Mientras esté en el lugar, compare las distintas calabazas. Anime al niñoa su hijo a que le diga lo que dos calabazas tienen en común y que señale las diferencias que puede identificar. Utilice las descripciones de su hijo para inspirarse y buscar juntos la calabaza más grande y la más pequeña. También pueden buscar la calabaza más irregular, la más plana, la más alta, la más redonda, etc. Muestre entusiasmo a medida que sigue el ejemplo de su hijo y descubran las calabazas juntos.

Assessment—What to Look For

- **Can children describe pumpkins based on their observations?**
  (using senses to gather information, identifying properties, using complex patterns of speech)
- **Can children compare characteristics of two or more pumpkins or other objects?**
  (identifying properties, comparing properties, classifying)
- **Can children measure and compare measurements?**
  (measuring, comparing size, using computational thinking)
- **Can children record (draw or write) observations and contribute to class discussion?**
  (using new or complex vocabulary, documenting and reporting findings, discussing scientific concepts, listening to and understanding speech)
- **Can children explain their reasoning?**
  (using complex speech patterns, constructing explanations, explaining based on evidence)

Standards

Head Start Early Learning Outcomes Framework
P-SCI 1. Child observes and describes observable phenomena (objects, materials, organisms, events, etc.). "Makes increasingly complex observations of objects, materials, organisms, and events. Provides greater detail in descriptions. Represents observable phenomena in more complex ways, such as pictures that include more detail."
Next Generation Science Standards
Science and Engineering Practice: Planning and carrying out investigations. "Make observations (firsthand or from media) and/or measurements to collect data that can be used to make comparisons."
Common Core State Standards for Mathematics
1.MD.A.1. Measure lengths indirectly and by iterating length units. "Order three objects by length; compare the lengths of two objects indirectly by using a third object."
Common Core State Standards for English Language Arts
SL.K.4. Presentation of knowledge and ideas. "Describe familiar people, places, things, and events and, with prompting and support, provide additional detail."

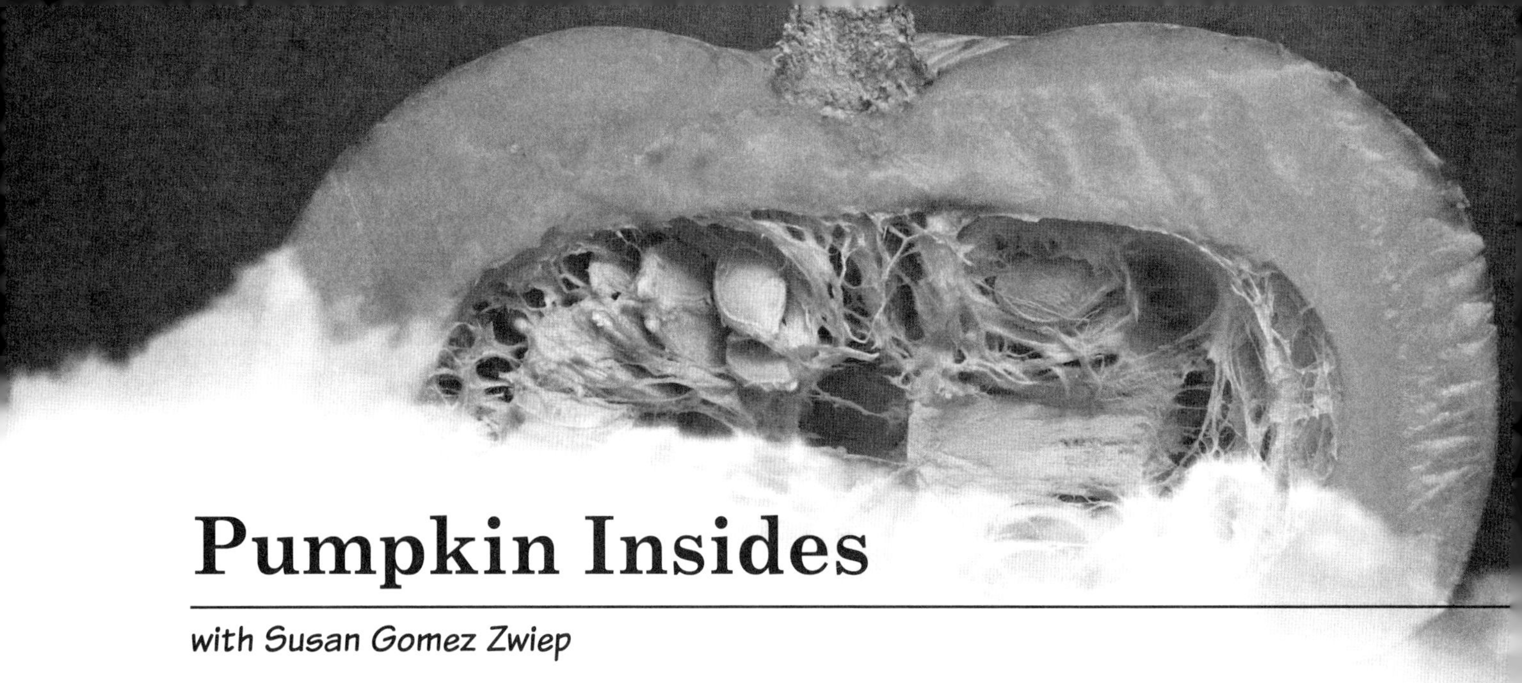

Pumpkin Insides

with Susan Gomez Zwiep

Lesson: Observing, describing, and comparing pumpkin insides

Learning Objectives: Children will observe, describe, and compare the different parts found inside a pumpkin.

Materials: Multiple pumpkins of various shapes, sizes and color; chart paper; writing and drawing materials; one strip of paper for each pumpkin (each strip should be about 2 inches wide and from 2 to 3 feet in length (as long as the circumference of the biggest pumpkin)

Safety: Although rare, food allergies to pumpkin, including pumpkin seeds, are possible. Find out if any of your children have allergies before using materials in this lesson and remind yourself of the symptoms of allergic reactions in children. Also, be aware of seeds as possible choking hazards. Remind children not to eat any food being used in the activity and not to put their fingers in their mouth or nose while handling plant material. Make sure children thoroughly wash their hands with soap and water after the activity

Teacher Content Background: A pumpkin is the fruit of a pumpkin plant. These fruits have several layers. The pumpkin "skin" or "rind" is the thin, shiny, often orange, outer layer of a pumpkin. It is a protective layer that keeps insects and disease out of the fruit. This part of the pumpkin is not eaten. The inside of the pumpkin, just below the skin, is the "pulp" or "flesh" of the pumpkin. This thick layer is the part of the pumpkin that can be cooked and eaten and is included in recipes for pies, cookies, soups, and many other delicious foods. Inside the pulp of a pumpkin is a goopy mixture of fibrous strands and seeds. The seeds each contain an embryo, a tiny plant. With the right combination of water and oxygen within an appropriate temperature range, the seed will germinate, and a tiny pumpkin plant within will emerge. (See *Sprouting Seeds*, p. 194, for more information about germination.) Seeds contain a temporary food supply that allows the embryo to live, germinate, and grow until the plant can photosynthesize and produce its own food. This temporary food supply is called the endosperm and is why seeds and nuts are a nutritious snack for us to eat. A single pumpkin might contain hundreds of seeds.

Science terms that may be helpful for teachers to know during this lesson include *investigate, observe,* and *pumpkin.*

Procedure

[Note: Although pumpkins are used here, many of these activities can be completed with any fruit. The advantage of pumpkins is their large size, making them easier for children to manipulate and observe. To prepare for this lesson, you will have to cut several pumpkins in half from top to bottom. Do this cutting carefully and in an area away from children. Also, *Pumpkin Outsides*, p. 116, can serve as an introduction to this lesson.]

Getting Started

Initial Explanation: Show children a pumpkin and ask children what they think might be inside. After this opening conversation, give each child a piece of paper with a pumpkin outline drawn on it and ask them to draw what they think is inside the pumpkin. Once children have completed their drawings, encourage them to describe what they drew. As children describe their drawings, write the different words they use to serve as labels for what they have drawn in their pumpkin outline. Use whichever words the children use to describe their thinking such as *hair, seeds, string, gooey stuff,* and so on. Once children are finished, have several share what they drew inside their pumpkin. Use questions such as "Have you ever seen the inside of a pumpkin? How did you know that those things are inside a pumpkin? Do you think each pumpkin has the same stuff inside or are they different?" to encourage children to describe what they already know about pumpkins.

Introduction: Before distributing pumpkins for the lesson, discuss with children how to observe the characteristics of an object. Together, come up with a list about what to look for when observing. This list may include determining what an object looks like (shape, size, color), feels like (smooth, rough, bumpy), sounds like (loud, banging), smells like (minty, smoky), and tastes like (sour, sweet). Review the five senses and, for each item on the list, draw the sense organ that corresponds with the descriptions (e.g., draw an eye for the "looks like").

Investigating

Observing: Tell children that they will now investigate the different parts of the inside of the pumpkin. Give each small group of children a pumpkin half to investigate. Tell children to observe the inside of the pumpkin using their sense of sight only—don't let them pull things out of the pumpkin just yet. After the children have had time to make some observations, tell them that they will now be able to touch and smell the different parts of the pumpkin. [Note: There is a tendency for children to start pulling things apart without careful observation. Caution against initial, frenzied pulling and digging. Instead, as you ask questions, model careful looking with minimal digging and tearing. Also, be sure to remind children that they will not be using their sense of taste to make observations.] Encourage children as they observe by asking, "What does it look like? Feel like? Are there parts that feel different? Sticky, rough, smooth? Is the inside of the pumpkin mostly full or mostly empty? What kinds of different parts do you see inside the pumpkin? Was there more of one part compared to the other parts?" After these initial observations, give children a new pumpkin outline and encourage them to draw what they observed inside the pumpkin.

Making Sense

Describing Findings: Engage children in a group discussion of what they observed inside the pumpkins. For each part suggested by the children, discuss their observations and ask children how you might draw each of the pumpkin parts on the chart paper. As you discuss each part of the pumpkin, try to get the children to decide what each of the different parts should be called. For example, the children might identify seeds, stringy stuff, or skin—it's fine if their descriptive terms don't quite match the conventional ones. Write each of the decided names on the chart paper, drawing lines between the names and the parts in your drawing. When all the parts have been described and named, review them and ask children to compare your drawing to what they have drawn on their own papers. If children are interested, allow them to add to or edit their drawings as they see fit and encourage children to describe the changes they are making. If children choose to, they may add labels to their drawings (see *Writing Connections*).

What's Next?

Extension Activity 1

Select and slice in half other types of fruits. [Note: As always, be purposeful in your selection of materials to present to children. Include some fruits that are similar to pumpkins (e.g., squash and melons) and some that are different (e.g., oranges and avocados), but be sure to avoid "seedless" varieties of fruits.] Ask children to compare these types of fruits to their pumpkins. Direct their attention to the different parts of these fruits — comparing the rinds, flesh, and seeds. Children should observe that some fruits are similar to pumpkins and some are very different. Depending on the fruits selected, children may discover differences, such as that some skins are smooth and some are bumpy, some pulp is hard and some is soft, and some seeds are flat and some are round. (For additional fruit explorations, see *Seeds in Our Food*, p. 81.)

Extension Activity 2

Your children can germinate pumpkin seeds and even grow pumpkin plants. Use seeds from your pumpkins and follow the procedures in the *Sprouting Seeds* lesson and extension activities on page 106. Just as they did with lima beans, children can soak, dissect, and germinate their pumpkin seeds. With lima beans, children compared the germination of soaked seeds to unsoaked seeds; with pumpkins, they can compare the germination of seeds from different varieties of pumpkin. Let children decide which tests they're interested in. Do seeds from orange pumpkins germinate faster than those from white pumpkins? Bumpy versus smooth? Large versus small? Together with your children, plan, carry out, and compare data from these experiments.

Integration to Other Content Areas

Reading Connections

The activities in this lesson include opportunities to draw children's attention to print and to use print in meaningful ways that support children's development of emergent literacy skills. These include describing and drawing pumpkins and comparing pumpkins to other fruits. Fill your classroom library with both fiction and nonfiction books about pumpkins for children to explore. Entertaining and informative narrative texts about pumpkins include *Ready for Pumpkins.* by Kate Duke (2012), *Pumpkin Pumpkin* by Jeanne Titherington (1990), and *From Seed to Pumpkin* by Wendy Pfeffer (2015). Informational books such as *Seed, Sprout, Pumpkin, Pie* by Jill Esbaum (2009); *See It Grow: Pumpkin* by Jackie Lee (2016); and *Fall Pumpkins: Orange and Plump* by Martha Rustad (2011) will help you and your children learn more about pumpkin plants and their fruits. As you read these books, fiction or nonfiction, encourage children to talk about what they know about pumpkins and to share any new questions they have or insights they learned from the books.

Counting seeds in a pumpkin

Writing Connections

You can further support children's writing development by encouraging children to add labels to their observational pumpkin drawings. Provide each child with labels for the different parts of a pumpkin (e.g., rind, pulp, fibers, and seeds). Help children use letter-sound recognition as they identify each label and apply it to their drawing with questions such as "The word *pulp* starts with which letter? Which of these labels is the word *pulp*? Where on your drawing should you put the *pulp* label?" Adding labels to their observations will engage children in scientific communication and help children to understand that print contains meaning. For creative writing, ask children to bring from home a pumpkin recipe written by an adult on an index card. Glue the cards to the bottom of a sheet of paper; with your assistance children can write their name and write or trace the name of their recipe at the top of the page. Compile the recipes into a book. Children can "read" the recipes to ensure that *pumpkin* is included in each one and can share the recipes with their families.

Math Connections

Share the book, *How Many Seeds in a Pumpkin?* by Margaret McNamara (2007) with children, but stop reading when Mr. Tiffin, the teacher in the story, says, "Tomorrow we will find out the answer to our question." In this book, a class of children investigate the number of seeds in different pumpkins. Rather than reading the answer in the book, use this story as inspiration for conducting your own counts of pumpkin seeds. Just like in the story, give children a large, a medium, and a small pumpkin and have children excavate the seeds from each. With your help, children can count the seeds and compare the results. As an alternative to counting, children can weigh the seeds to determine which pumpkins have the most. And many other pumpkin-themed books can be used to spark discussions about shapes or counting. Try *Pumpkin Fever* by Charnan Simon (2011), *Pick a Circle, Gather Squares* by Felicia Sanzari Chernesky (2013), *Ten Orange Pumpkins* by Stephen Savage (2013) or *Pumpkin Countdown* by Joan Holub (2012).

Other Connections

Child's Life Connection

The outsides of pumpkins are hard and dry, but the insides are stringy and slimy. Children can take their understanding that the insides of some objects can have characteristics that are very different than what the objects look like on the outside and apply this understanding to objects in the classroom and in their homes. Let children explore and identify objects such as pencil boxes, backpacks, lunch bags, and cabinets. Encourage children to describe the outsides of these objects as well as the characteristics of the different parts inside that cannot be seen from the outside. Children can explore things at home and bring drawings or photographs of the objects back to school to share with the rest of the class.

Center Connections

In the **art center**, children can experiment with color mixing. Provide only primary color paints (blue, red, yellow) and have children discover which two colors, when mixed, can make orange and which two can create green. Provide a pumpkin outline and let children use their mixed primary colors to paint an orange pumpkin with a green stem. At the **sensory table**, of course put out pumpkins! Remove the tops from several pumpkins and allow children to reach in and explore. Encourage children to use descriptive language as they explain how the inside of the pumpkin feels. Prompt children to also describe how the outside of the pumpkin feels and to compare this with the inside. For **dramatic play**, children can play school, pretending to be teachers and children learning about pumpkins. Put out chairs for children and an easel and pointer for the teacher and supply different materials from your lesson, including pictures of pumpkin plants, samples of pumpkins and pumpkin seeds, and vocabulary words for the teacher to show his or her children.

Family Activities

At school your child has been investigating the insides of pumpkins. Ask your child about the pumpkin activities she or he has been doing and about the different parts of a pumpkin. If you can, purchase a pumpkin and a few other fruits from a store. (Be sure the fruits you purchase have seeds.) Carefully, and away from your child, cut the pumpkin and other fruits in half. Together, you and your child can compare the different fruits to the pumpkin. Fruits have different layers: skin (the outside layer), pulp (the often juicy layer within the skin), and seeds. Let your child be the teacher. As he or she shows you how the fruits are similar to and different from the pumpkin, be an enthusiastic and inquisitive learner. Depending on the fruits selected, you may discover that some skins are smooth and some are bumpy, some pulp is hard and some is soft, and some seeds are flat and some are round.

Actividades Familiares

En la escuela, su hijo ha estado investigando el interior de las calabazas. Pregúntele a su hijo sobre las actividades que ha estado llevando a cabo en torno a las calabazas y las distintas partes de ellas. Si puede, compre una calabaza y algunas otras frutas en una tienda. (Asegúrese de que las frutas que compre tengan semillas). Con cuidado y lejos de su hijo, corte la calabaza y las otras frutas por la mitad. Juntos, usted y su hijo pueden comparar las diferentes frutas con la calabaza. Las frutas tienen capas distintas: la cáscara (la capa exterior), la pulpa (la capa a menudo jugosa dentro de la piel) y las semillas. Deje que su hijo sea el maestro. Muestre entusiasmo y curiosidad por aprender a medida que su hijo le muestra las semejanzas y diferencias entre la calabaza y las otras frutas. Dependiendo de las frutas que haya seleccionado, puede descubrir que algunas cáscaras son suaves y algunas irregulares, algunas pulpas son duras y otras suaves y algunas semillas son planas y otras redondas.

Assessment—What to Look For

- **Can children describe the insides of the pumpkins and other objects based on their observations?** (using senses to gather information, identifying properties, using new or complex vocabulary)
- **Can children compare characteristics of two or more pumpkins or other objects?** (comparing properties, explaining based on evidence, using new or complex vocabulary)
- **Can children record (draw or write) observations and contribute to class discussion?** (using new or complex vocabulary, documenting and reporting findings, discussing scientific concepts, listening to and understanding speech)

Standards

Head Start Early Learning Outcomes Framework
P-SCI 1. Child observes and describes observable phenomena (objects, materials, organisms, events, etc.). "Makes increasingly complex observations of objects, materials, organisms, and events. Provides greater detail in descriptions. Represents observable phenomena in more complex ways, such as pictures that include more detail."
Next Generation Science Standards
Science and Engineering Practice: Planning and carrying out investigations. "Make observations (firsthand or from media) and/or measurements to collect data that can be used to make comparisons. Make predictions based on prior experiences."
Common Core State Standards for Mathematics
K.MD.A.2. Describe and compare measurable attributes. "Directly compare two objects with a measurable attribute in common, to see which object has 'more of'/'less of' the attribute, and describe the difference."
Common Core State Standards for English Language Arts
W.K.2. Text types and purposes. "Use a combination of drawing, dictating, and writing to compose informative/explanatory texts in which they name what they are writing about and supply some information about the topic."

4

Nature Walks

The benefits of outdoor learning are well documented and include increases in children's motivation and engagement, school performance, connection to nature and their community, physical activity, creativity, and problem solving, among many other positive outcomes. Fortunately, you and your children don't need a forest to reap these benefits. Nature can come in many forms and can be found in many places—a local garden, a neighborhood park … even a single tree can serve as a context for learning about and connecting to nature. As you explore nature with your children, the lessons in this chapter will serve to provide opportunities to focus your children's explorations and to help them to discover a variety of living things and forms found in nature as they investigate birds, trees, animals, leaves, and much more.

To maximize your children's outdoor learning experiences, keep in mind some basic guidelines for conducting field trips. Scout out your trip before taking children there—look carefully for safety issues, learning opportunities, and possibilities for your children to freely explore. Have a clear goal in mind for your trip, and identify or design specific activities to help reach this goal. Recruit parents to help, and train parents to be attentive to children's safety and to be effective guides and inquisitive listeners in children's explorations. During your time outdoors, find a balance between open exploration (as children follow what interests them at the site) and teacher facilitation (as you or other adults direct children to observe and consider items of interest that they might have otherwise missed). After the trip, be sure to follow up with discussions and activities back in the classroom. And throughout all of your teaching, let children's interests and children's decisions shape the direction and scope of the learning experiences you offer.

Animal Walk

with Angelica Gunderson

Lesson: Observing, describing, and counting animals

Learning Objectives: Children will go on a nature walk at a local park (or around the school grounds) to observe and count the types of animals that live there and to learn about animals' basic needs.

Materials: Chart paper, markers, data sheet and pencils or crayons to record observed animals, camera, binoculars

Safety: This lesson will involve being outside. As with all field trips, scout out the area before you take children there; make sure the area is safe and free from hazards. Be aware of children in your class that may have allergies (pollen, bees, etc.) and plan accordingly. Be careful to stay away from poisonous plants. Also, be mindful of roots, obstructions, and other hazards along your path outside. It will be helpful to have additional adults to help supervise and interact with children while you are outside.

Teacher Content Background: When asked to consider the animals in the area, your children might first think of pets and other domesticated animals. However, your neighborhood—whether urban, suburban, or rural—is home for many different wild animals. An animal's ability to live in a particular place, whether in a city or in the wild, depends on its ability to meet a few basic needs: oxygen, water, food, and shelter. Oxygen is essential for nearly all animal life. Among other roles, oxygen is required at the cellular level to release energy from food (glucose). We most often think of oxygen as a component of air, but it is also dissolved in water. Water is vital to animal life for many, many reasons—which is not surprising, considering that an animal's body is 50% to 95% water. The roles of water include regulating body temperature, digestion, excretion, and the absorption and transportation of nutrients. Food provides sources of energy and nutrients for animal metabolism, growth, and movement. Most species of animals have adapted to utilize specific types of foods. These specific dietary needs help to determine appropriate habitats for different animals. Shelter provides animals with protection from the environment and from predators. To limit competition between species, most animals have specific needs for shelter. For example, different species of birds that nest in trees will make their nests from different materials and place their nests in different types of trees and at different locations within a tree. The availability of water and different types of food and shelter help to determine the specific animals that can be found in a given area, including neighborhood parks.

Science terms that may be helpful for teachers to know during this lesson include *observe, animal,* and *graph*.

Procedure

[Note: For this lesson, focus on larger, more conspicuous animals such as lizards, squirrels, birds, butterflies, rabbits, etc. If animals are hard to find, encourage children to search for clues (birds' nests, spiderwebs, gopher holes, etc.) that might indicate the different types of animals that live in the area.]

Getting Started

Introduction: To begin this lesson, get children thinking about animals. Ask them, "What is an animal?" As children respond, they may initially suggest examples of animals. Welcome these responses, but eventually direct the conversation toward the characteristics of animals, asking children questions such as "Is this table an animal? What about this plant; is it an animal? Why not?" During the discussion, children will point out that animals are living. Follow up on this, asking children for their ideas about what animals need to live. Expect a range of answers and list each child's suggestions. Review the list with children and have them decide which two to four factors are most important. [Note: It's fine if they don't select air, food, water, and shelter as the most important needs. The goal is for young learners to understand that animals live in places that meet their basic needs. Throughout the *Investigating* portion of the lesson, ask children how each animal is able to meet its needs.]

Initial Explanation: Let children know that during their science lesson, they will be looking for animals. Ask children, "What are some of the animals that we might find in our neighborhood?" Record each animal using words and picture support. For each animal listed, ask children, "Where are some places in our neighborhood where we may be able to find this animal? Why would that animal live there?" Have children share their ideas and record these next to each animal, again using words and picture support.

Investigating

[Note: Children can also map the locations of the animals they find—see *Finding Weeds*, p. 151, for a description of mapping with children.]

Observing: Take the children outside to the location of your planned walk. [Note: Be sure to have scouted out your walk ahead of time to identify specific places with lots of animals as well as possible hazards to avoid.] At several places along your walk, have children sit and look and listen for animals. When an animal is located, ask children if they know the name of the animal and, if needed, tell children the name. Let children share what they know and notice about the animal. Then use questions that encourage children to make more detailed observations about both the appearance (e.g., "What colors does the [observed animal] have?" "What other animal does the [observed animal] look like?" "How is the [observed animal]'s body different from yours?") and the behavior (e.g., "Watch the way the [observed animal] walks; can you imitate the [observed animal]'s walk?" "Where do you think the [observed animal] sleeps? How does it make its home?" "What do you think the [observed animal] eats? How can you tell?") of the animal. Also ask the children about how the animal is meeting its most important needs. The children can use the binoculars to get a closer look and cameras to take pictures of the animals they see.

Recording Data: Give children data sheets with pictures and names of the different animals discussed during the *Getting Started* portion of the lesson. (See Figure 4.1.) Be sure to leave space for children to add additional animals not discussed earlier. On their data sheets, children can record tally marks for each animal they see. Once children have observed and tallied the animals at one location, move to another; let children's interest guide you in determining the number of stops to make.

Figure 4.1

Animal Data Sheet

Squirrels	Lizards	Birds
Insert picture of squirrel here	Insert picture of lizard here	Insert picture of bird here
How many did you see?	*How many did you see?*	*How many did you see?*
Butterflies	**Rabbits**	
Insert picture of butterfly here	Insert picture of rabbit here	
How many did you see?	*How many did you see?*	*How many did you see?*

A full-size version of this figure is available at *www.nsta.org/startlifesci.*

Making Sense

Describing Findings: Back in the classroom, review the ideas generated the *Getting Started* portion of the lesson. Were children able to see all of the expected animals? What unexpected animals did they see? For each animal, prompt children to discuss features and behaviors they noticed. As you did during the animal walk, use questions to help children to think about, describe, and compare the different animals and to discuss how animals are able to meet their most important needs. Animals can only live in places that provide their important needs; model for children and help them to express reasoning that demonstrates this idea. For example, "Lizards can live in the park because the park provides lizards with [important needs] (e.g., food, water, and a place to sleep)."

Direct children's attention to their data sheets, asking them to count how many of each animal they saw and to decide which type of animal was the most common. There will be some discrepancies in counts, so create a class data sheet together on the whiteboard or on a piece of chart paper. Have individual children report their number for an animal and use these data to create a bar graph together (see *Math Connections*).

[Note: If your children finish this lesson eager to learn more about local animals, *Critter Camouflage*, p. 51, can make for an interesting next lesson.]

What's Next?

Extension Activity

You can extend children's learning by repeating this lesson. Repeating the lesson in a new location will allow children to compare the types and numbers of animals found in this location with those found during your initial class trip. Choose a location that is unlike your local park (i.e., has fewer trees and bushes). Collecting and comparing data about the animals found in parks with those found in parking lots, playgrounds, or neighborhoods will provide children an opportunity to describe how these two areas differ and to further discuss the basic needs of animals. You can encourage this discussion by asking about the different animals and different resources found at each site. For example, "Where in the park did we find squirrels? How do trees help the squirrels? Why do you think we didn't find any squirrels in the parking lot?" Another useful extension is to repeat the investigation in the same location but at different times of the year. During different seasons, different animals may be present. Comparing data sheets from different times of the year will give children an opportunity to analyze data and to speculate about the behaviors of different animals and how animals are affected by the seasons.

Integration to Other Content Areas

Reading Connections

We share our neighborhoods with many different animals. Many books can inform children's understanding of, and inspire their searches for, these animals. Books such as *Coyote Moon* by Maria Gianferrari (2016); *Wild in the City* by Jan Thornhill (1996); *Squirrels Leap, Squirrels Sleep* by April Pulley Sayre (2016); and *City Animals* by Simms Taback (2009) can expand children's understanding of the different animals that may be living in their neighborhood. Many other books, including *Animal Homes* by Debbie Martin (1999), *My Very First Book of Animal Homes* by Eric Carle (2007), *Peek Inside Animal Homes* by Anna Milbourne (2014), and *Animal Homes* by Ann Squire (2002), can be used to prompt further discussions about animal needs and how their homes help provide these.

Writing Connections

When children are describing and comparing observations and recording data, they are developing their literary and language skills. To further develop their emergent literacy, children can use the data and pictures collected during the lesson to create a counting book that shows the number of each animal observed during the class trip. Write or print out the title of the book (e.g., *Animals in Our Park*) for children to trace and print out sheets with several small copies of each picture. Children can cut out and glue the appropriate number of copies on one page of the book. On the opposite page, children can write (with support as needed) the number and the name of the animal (e.g., "four squirrels") as they complete the sentence "____ _____ live in the park." Encourage children to read the book that they wrote to each other and to family members after school.

Math Connections

This lesson presents several opportunities for children to practice collecting, representing, and interpreting data. Using graph paper and data collected during the animal walk, children can create a bar graph. Have children label the bottom of the graph with the pictures and names of the different types of animals they have seen. Then children can count tally marks on their data sheets and color boxes or add stickers to represent each individual animal seen. Children can analyze their graphs to determine the types of animals that are most and least abundant. Children can also count and create graphs that compare different animal attributes, such as animals that fly and those that do not, animals that can be pets and those that cannot, animals that are brown and those that are not, and so on.

Other Connections

Child's Life Connection

The animals that children know and love best are their pets. Ask children to share information about the type of pet they have or would like to have. Prompt them to describe their pets' needs, explaining what their pet eats and does. If children don't have pets, ask them to pick an animal, possibly an animal they saw during the lesson, that they would like to have for a pet and to explain why that animal would make a good pet. Help children to consider different animals. Give children information about the shelter and food needs of an animal and prompt children to consider if the animal would make a good pet or not (e.g., "Bats live in caves and eat flying insects. Would a bat be a good pet for our classroom? Why not?")

Center Connections

In the **art center**, children can use the pictures of animals as models for artistic renditions. Provide many different materials and encourage children to use two different media in their art (e.g., paint and feathers to create a bird, cotton balls and tissue paper to make a rabbit). At the **sensory table**, put out several different swatches of material; include material that is soft, furry, rough, or bumpy. Children can explore the different textures with their hands. Encourage them to describe the way each material feels and to discuss which material they think would feel most like each of the different animals they saw during the lesson. Children can turn the **dramatic play** area into a park. Put out a few trees or bushes (potted or plastic) and provide animals (stuffed animals or pictures of animals) and different materials the animals would need to survive (sticks for nests, acorns to eat, etc.). Children can populate their park with different animals and act out different animal interactions. Encourage children to pretend to be mommy or daddy animals that are responsible for getting food and shelter for their babies.

Family Activities

At school, your child has been looking for and learning about animals that live in our neighborhood. You can help your child continue to look for local wildlife. Go to a park and take a quiet walk with your child to look and listen for different animals. When you see or hear an animal, let your child be the expert and tell you what he or she knows about it. Ask your child what he or she thinks the animal is doing, what it eats, and where it lives. (Don't worry if your child isn't exactly right; the process of observing animals and sharing and explaining his or her observations with you is what's most important.) During a similar walk at school, we counted the different animals we saw; your child may want to do the same with you on your walk. Keep track of how many of each type of animal you see and, after your walk, your child can compare and tell you about the different numbers of animals.

Actividades Familiares

En la escuela, su hijo ha estado buscando y aprendiendo sobre los animales que viven en el vecindario. Puede ayudar a su hijo a seguir buscando fauna local. Llévelo a un parque y dé un paseo tranquilo con su hijo para ver y escuchar a los distintos animales. Cuando vea o escuche un animal, deje que su hijo sea el experto y que le diga acerca de lo que sabe. Pregúntele qué piensa que el animal está haciendo, lo que come y dónde vive. (No se preocupe si su hijo no está exactamente en lo correcto, el proceso de observar animales, compartir y explicar sus observaciones con usted es lo más importante.) Durante un paseo similar en la escuela, contamos los diferentes animales que vimos; su hijo puede querer hacer lo mismo con usted durante su paseo. Lleve un registro de cuántos ejemplares de cada tipos de animales ven y, después de un paseo, su hijo puede comparar e y le informarleá sobre las cantidades de los distintos animales.

Assessment—What to Look For

- **Can children make predictions about the types of animals that may live in their neighborhood?**
  (making predictions, making inferences, using complex patterns of speech)
- **Can children describe the features and behaviors of the animals they find?**
  (identifying properties, comparing properties, using new or complex vocabulary)
- **Can children identify one or more animal needs?**
  (identifying properties, constructing explanations, using new or complex vocabulary)
- **Can children record (draw or write) observations and contribute to class discussion?**
  (using new or complex vocabulary, documenting and reporting findings, discussing scientific concepts, listening to and understanding speech)

Standards

Head Start Early Learning Outcomes Framework
P-SCI 1. Child observes and describes observable phenomena (objects, materials, organisms, events, etc.). "Makes increasingly complex observations of objects, materials, organisms, and events. Provides greater detail in descriptions. Represents observable phenomena in more complex ways, such as pictures that include more detail."
Next Generation Science Standards
Science and Engineering Practice: Analyzing and interpreting data. "Use observations (firsthand or from media) to describe patterns and/or relationships in the natural and design world(s) in order to answer scientific questions and solve problems."
Common Core State Standards for Mathematics
1.MD.C.4. Represent and interpret data. "Organize, represent, and interpret data with up to three categories; ask and answer questions about the total number of data points, how many in each category, and how many more or less are in one category than in another."
Common Core State Standards for English Language Arts
SL.K.1. Comprehension and collaboration. "Participate in collaborative conversations with diverse partners about kindergarten topics and texts with peers and adults in small and larger groups."

Looking for Birds

with Kristin Straits

Lesson: Observing birds in the environment

Learning Objectives: Children will observe and gain an appreciation for the diversity of birds in their local area. Children will compare different birds to identify the characteristics they have in common.

Materials: Drawing and writing materials, camera, photographs or other visuals of different local birds

Safety: This lesson will involve being outside. As with all field trips, scout out the area before you take children there; make sure the area is safe and free from hazards. Be aware of children in your class that may have allergies (pollen, bees, etc.) and plan accordingly. Be careful to stay away from poisonous plants such as poison ivy. Also, be mindful of roots, obstructions, and other hazards along your path outside. It will be helpful to have additional adults to help supervise and interact with children while you are all outside. While searching for birds, remember that although beautiful and often relatively small, birds are wildlife; it's best to avoid contact with birds, especially if they appear to be sick or injured. Be sure none of your children are allergic to feathers before using feathers in the classroom.

Teacher Content Background: There are more than 700 different species of birds that live year-round in North America or at least occasionally visit the area. Across these 700+ species there is a great deal of variety in bird behavior and appearance; consider the tiny ruby-throated hummingbird and the imposing golden eagle. It's this variety that has made birding (i.e., the purposeful identification and study of birds) one of the most popular outdoor hobbies in the United States. Although many of our common birds are fairly conspicuous (e.g., blue jay, crow), most are somewhat secretive and harder to see (e.g., many warblers). When looking for these less obvious birds, it is often helpful to focus on looking for movement. Perhaps more important than looking is listening; birds are often heard before they are seen. In fact, the songs and calls of birds are often more distinctive than the pattern or color of their plumage. Try to learn both the look and sound of the most common birds in your area. Once you've identified a bird, you can then really observe it to learn much about the bird's behavior. There are many inexpensive field guides, as well as online resources, to assist in this endeavor.

Science terms that may be helpful for teachers to know during this lesson include *observe*, *behavior*, *bill*, *feather*, *tail*, and *wings*.

Procedure

[Note: Much like *Animal Walk*, p. 131. and many other lessons presented here, this lesson can be repeated during different times of the year to investigate how seasons affect birds.]

Getting Started

Prior Knowledge: Begin a discussion about birds by asking children where they have seen birds. You can encourage children to respond with guiding questions such as "Have you seen birds in the sky? On the water? In trees?" Let the conversation grow, using open-ended questions when needed, and allow children to share their ideas about birds, different kinds of birds, different things that birds do, and any questions they have about birds.

Prompting Questions: After children have had a chance to share with each other, ask, "What about here at school—are there birds here?" Discuss the birds children have seen at school. The names of the birds aren't important here—instead, emphasize children's description of birds. "How big was the bird?" "Where did you see the bird?" "What color was its feathers?" "What was the bird doing?" Tell children that birds are one of the many animals that we share our environment with, and today they're going to learn more about the birds that can be seen right here at school. Encourage children to predict which birds they'll find as you explore your school.

Investigating

Observing: Begin your investigation of birds by taking your young children on a "listening walk." Ask them to listen very carefully for sounds of any type. What are they hearing? Depending on where you are and the events going on around you, the range of sounds your children will hear may be broad indeed. Among all that is being heard, are there any bird sounds? If so, ask the children to describe them and even to try to imitate what they are hearing. [Note: If the area around your school seems to lack birds and their sounds, consider going to a nearby park. Places with trees, lawns, and shrubs are more likely to have birds nearby. And, remember: so-called urban birds (e.g., pigeons, crows, house sparrows) are much more widespread in city neighborhoods than we sometimes think.]

Observing and Comparing: Encourage your children to become careful observers. Be on the lookout for different kinds of birds and, along the way, ask probing questions to encourage more careful observation. Consider questions that direct children to observe, compare, and think about birds' appearances (e.g., "What colors do you see in that bird's feathers?" "How are those two birds alike?" "What can you tell me about

that bird's beak? Do the beaks of those two birds look alike or different?" "How is the bird's foot different from yours?"), sounds (e.g., "Can you hear that bird? What does it sound like?" "Can you imitate the sounds that crow is making?" "How are the sounds of the crow different from those of that robin?" "Which of the birds that you've seen makes sounds you like best? What do you like about those sounds?"), and behaviors (e.g., "Watch the way that pigeon walks. Can you imitate the pigeon's walk?" "Does the sparrow move the same way the pigeon does? Tell me about the way they move." "How does that bird eat? How is your way of eating different from that of the bird?"). Remember that the intention is to enhance children's awareness and curiosity about birds in our everyday world. Strive to promote children's fascination with nature and engage in friendly conversations with your children about birds and birds' appearances, sounds, and behaviors.

Documenting: Provide children with a camera to take photographs of the different birds they see. [Note: Be sure that you take photographs as well, to ensure that the class has photographs that detail each bird.] Back in the classroom, children can use these photographs, along with the observations they remember, as the basis for creating bird drawings. Displaying photographs alongside the children's drawings on a bulletin board will inspire conversations about the birds children have seen and these pictures will help in the *Making Sense* section below.

Making Sense

Identifying Patterns: Review bird photos and drawings and encourage children to discuss similarities and differences they notice. Show pictures of two different birds and ask children, "How are these two birds similar?" and record children's responses. Repeat this for several different birds that were seen during the *Investigating* part of the lesson. As you review their responses, try to build children's understanding of characteristics common to all birds. "The crow had wings. Did the pigeon have wings? What about the robin? Do you think *all* birds have wings?" Repeat this for several characteristics (e.g., colored feathers, bills, etc.) [Note: Be aware that based on the birds observed in your schoolyard, children may reasonably want to suggest that all birds fly. Not all birds fly (e.g., ostrich and penguin), but all birds *move*. Encourage children to recall birds that walk, hop, swim, and fly and emphasize that although there are many different ways that birds can move, all birds move.]

What's Next?

Extension Activity 1

After the walk, let children talk about their birds. Ask children if they would like more birds to visit their schoolyard. Ask what they think they could do so that more birds would want to visit. Listen to their responses, and then tell children that birds use a lot of energy when they fly, so they need to eat quite often. Ask them what they think birds might like to eat. Make bird feeders, put out a few samples of food, and watch for which foods the birds seem to like. [Note: *Feeding Birds*, p. 65, provides additional information and a possible follow-up lesson to this activity.]

Extension Activity 2

Plant a bird-friendly garden. Gardens in the schoolyard not only offer children wonderful learning opportunities but also help to encourage birds to visit your school. Some plants are more effective in attracting birds than others, so as you plan a school garden, include the birds in your planning. Think beyond the flowers, fruits, and vegetables that people typically include in their gardens and instead choose plants with the flowers, fruits, and seeds that your local birds would prefer. Incorporating a garden into your classroom exploration of birds provides wonderful opportunities to engage children in conversations about what birds need in order to survive. In general, birds seek out places that offer food, shelter, and water. Talking with children about birds and their basic needs provides an opportunity to remind children that these basic needs are shared by all living things. (See *Animal Walk*, p. 131, for an additional investigation about the basic needs of animals.)

Integration to Other Content Areas

Reading Connections

Children can learn more about birds with books such as *What Makes a Bird a Bird?* by May Garelick (1995), *About Birds: A Guide for Children* by Cathryn Sill (2013), and *Feathers: Not Just for Flying* by Melissa Stewart (2014) and can gain a greater appreciation of the diversity of bird life with books such as *Counting Is for the Birds* by Frank Mazzola (1997), *Have You Seen Birds?* by Joanne Oppenheim (1988), and *Bring on the Birds* by Susan Stockade (2013). Sharing this diversity may spark children's interest in looking for and identifying even more birds. Jim Arnosky's *Crinkleroot's 25 Birds Every Child Should Know* (1993) and *Crinkleroot's Guide to Knowing the Birds* (1992) are great starting points for our youngest learners. You can assist older children with more text-rich bird identification books such as *Bird Guide of North America* (2013) by Jonathon Alderfer and *Stokes Beginner's Guide to Birds* by Donald Stokes (1996). In addition to looking different, different types of birds make different, often distinctive sounds; books such as *Birdsongs* by Betsy Franco (2007), *Bird Talk* by Ann Jonas (1999), and *Birdsong* by Audrey Wood (1997) share different bird sounds and can encourage a great deal of squawking and quacking and kee-yawing from your children.

Writing Connections

Throughout the lesson, children have several opportunities to further develop their emergent literary and language skills, such as when children are making and recording (i.e., drawing) observations. Children can revisit their drawings or photographs, adding labels of important bird characteristics. This type of labeling helps children to understand that print contains meaning and that scientists use drawings and written words as tools to share what they have learned about our world. Birds can also serve as creative writing prompts for young children. Have your children make up a story about what it feels like to be a bird soaring high in the sky. Or begin a story with a robin's egg that is about to hatch, and ask children to describe what it's like for the baby robin to find its way out of the shell and into the outside world. Challenge younger learners to "write" the story with drawings to which you can add the words they dictate to you.

Math Connections

This lesson can be repeated, as different types and numbers of birds will be found on different days. As the children observe the birds near the school, help them to count the numbers of different birds and to keep track of this information in a variety of ways. Many times, numbers or tallies may be rather carelessly scribbled on paper. Encourage your children to be neat and consistent with regard to how their data are recorded, and encourage children to talk about helpful ways to organize their data. Whenever possible, use ideas that emerge from children. Although it

Figure 4.2

Bird Observation Data Sheet

[name of common bird #1]	How many did you see?	[name of common bird #2]	How many did you see?
[name of common bird #3]	How many did you see?	[name of common bird #4]	How many did you see?

A full-size version of this figure is available at *www.nsta.org/startlifesci.*

may seem more efficient to simply tell them what to do, their learning will be enhanced if you can encourage them to develop their own more organize way of doing things. (If, however, you'd prefer to provide children with a data sheet, consider the sample provided in Figure 4.2.) The data children collect can be tallied, counted, and graphed to determine which types of birds are most common near your school.

Other Connections

Child's Life Connection

Each state (as well as most countries, territories, and provinces) has an officially designated "state bird." For example, New York's official bird is the eastern bluebird, Texas has the northern mockingbird, and California has selected the california quail as its representative. Find out what your state's official bird is and display color pictures. Encourage children to look for their state bird in their neighborhood. A nice way of engaging children in conversations about states and their state birds is to invite them to talk about places they have been to or would like to go ("Let's find out the state bird of Alaska") and to think about relatives and friends living elsewhere ("I wonder what the state bird is where your grandmother lives?").

Center Connections

In the **art center,** children can use feathers in a variety of ways. If your supply of feathers is ample and varied, children can create feather collages by gluing specimens on colorful construction paper. Feathers can also be used as "paintbrushes" to create abstract patterns on paper. This works best if the paint is not too thick; you might allow children to experiment with paints of different consistencies. You can find feathers at most arts and crafts stores. At the **sensory table,** replace your usual sand or water with feathers. Let children explore feathers with their hands. Encourage them to notice the different textures within a feather, observing that there are parts of feathers that are soft and fluffy, smooth and hard, soft and firm, pointy, etc. For **dramatic play,** create or have children help you to create an aviary. Include many birds (stuffed animals or pictures), materials for feeding and caring for birds or creating bird nests, and bird costumes for children to wear (beaks, wings, etc). Include a sound track of various bird songs and have children imagine and act out being a bird. Children can walk like a pigeon, hop like a sparrow, waddle like a duck, and soar high in the sky like a hawk. They can act out different parts of a bird's life such as hatching out of an egg, learning to fly, and building a nest. Other children can play the role of bird-watcher, using "binoculars" (made from a paper towel roll) to watch the "birds." Children should have writing materials for drawing sketches of birds as bird watchers often do.

Family Activities

At school your child has been observing birds. You can help your child continue to look for birds. Go to a local park (or forest or beach), take a quiet walk with your child, and look and listen for birds. Once you find a few, observe what they are doing. Have your child help you count all the birds you see and remember their different colors. After the walk, encourage your child to share what she or he remembers seeing by asking questions such as "Where did we see most of the birds? In the trees? In the air? In the brush? Near or on the water? Why do you think there were so many birds there? We saw so many birds that had different colors! Can you remember some of the colors of the birds? Which ones did you like best?" Share what you remember about the birds you saw. Enjoy exploring nature with your child and make plans to do it again soon.

Actividades Familiares

En la escuela, su hijo ha estado observando aves. Puede ayudar a su hijo a seguir con su búsqueda de aves. Visite un parque local (o el bosque o la playa), dé un paseo tranquilo con su hijo y observen y escuchen a los pájaros.las aves Una vez que encuentre unos pocos, observe lo que están haciendo. Pídale a su hijo que le ayude a contar todas las aves que ven y que recuerde sus distintos colores. Después del paseo la caminata, anime a su hijo a compartir lo que recuerda haber visto, hágale preguntas tales como "¿Dónde vimos la mayor parte de las aves? ¿En los arboles? ¿En el aire? ¿En los arbustos? ¿Cerca o en el agua? ¿Por qué crees que hay tantas aves allí? ¡Vimos muchas aves que tenían diferentes colores! ¿Puedes recordar algunos de los colores de los pájaros las aves? ¿Cuáles te gustaron más?" Comparta lo que recuerda de las aves que vio. Disfrute el explorar la naturaleza con su hijo y haga planes para hacerlo de nuevo pronto.

Assessment—What to Look For

- **Can children make predictions about the types of animals that may live in their neighborhood?**
  (making predictions, making inferences, using complex patterns of speech)
- **Can children describe the features and behaviors of the animals they find?**
  (identifying properties, comparing properties, using new or complex vocabulary)
- **Can children identify one or more animal needs?**
  (identifying properties, constructing explanations, using new or complex vocabulary)
- **Can children record (draw or write) observations and contribute to class discussion?**
  (using new or complex vocabulary, documenting and reporting findings, discussing scientific concepts, listening to and understanding speech)

Standards

Head Start Early Learning Outcomes Framework
P-SCI 1. Child observes and describes observable phenomena (objects, materials, organisms, events, etc.). "Makes increasingly complex observations of objects, materials, organisms, and events. Provides greater detail in descriptions. Represents observable phenomena in more complex ways, such as pictures that include more detail."
Next Generation Science Standards
Science and Engineering Practice: Planning and carrying out investigations. "Make observations (firsthand or from media) and/or measurements to collect data that can be used to make comparisons. Use observations (firsthand or from media) to describe patterns and/or relationships in the natural and designed world(s) in order to answer scientific questions and solve problems."
Common Core State Standards for Mathematics
1.MD.C.4. Represent and interpret data. "Organize, represent, and interpret data with up to three categories; ask and answer questions about the total number of data points, how many in each category, and how many more or less are in one category than in another."
Common Core State Standards for English Language Arts
SL.K.1. Comprehension and collaboration. "Participate in collaborative conversations with diverse partners about kindergarten topics and texts with peers and adults in small and larger groups."

Nature Bracelets

with Emily Kraemer

Lesson: Exploring nature and the different structures of plants, especially flowers

Learning Objectives: Children will explore the outdoors to discover different objects in nature. Children will sort objects to learn about and identify different plant structures with particular attention to the parts of a flower.

Materials: Masking tape, magnifiers, several trays for sorting, chart paper or poster board

Safety: This lesson will involve being outside. As with all field trips, scout out the area before you take children there; make sure the area is safe and free from hazards. In particular, look carefully for poison ivy, poison oak, stinging nettle, and other plants that can cause severe skin irritations. Be aware of children in your class who may have allergies (pollen, bees, etc.) and plan accordingly. Also, be mindful of roots, obstructions, and other hazards along your path outside. It will be helpful to have additional adults to help supervise children while you are all outside. At the end this lesson, be sure children thoroughly wash their hands with soap and water.

Teacher Content Background: The possible items for children to collect for their nature bracelets are nearly endless. However, one of the most popular choices is to collect flowers and flower parts. A typical flower has four parts: the sepals (small, leaf-like structures that protect the flower bud before it blossoms), petals (typically large, colorful, leafy structures that attract pollinators), stamen (often numerous structures that produce and release pollen), and the pistil (the central structure of a flower—this contains ovules and becomes the fruit). Not all plants produce flowers and not all flowers are large and colorful, but all flowers function in reproduction. For plants, reproduction has two important events: first, pollination (the transfer of sperm-containing pollen from one flower to another) and second, fertilization (the union of sperm and ovule within the pistil). As plants are rooted in place, it's the transfer of pollen from one plant to another that is the greatest challenge. Generally, the two strategies for achieving this transfer among flowering plants are using the wind and using animals. The sepals and petals of wind-pollinated flowers (e.g., most grasses and many trees) are very small or completely absent to avoid interfering with pollen blown from or to the flower. The far more common adaptation is to use animals to transfer pollen. Many birds and bees are pollinators, but there are plants that use flies, bats, geckos, and even lemurs to transfer pollen. To entice potential pollinators, plants produce nectar and excess pollen and advertise these food sources using sweet aromas and brightly colored petals or sepals. (See *Smelling Plants*, p. 109, to read more about the different scents plants produce.)

Science terms that may be helpful for teachers to know during this lesson include *sorting*, *dissect*, *petal*, *leaf*, *flower*, *twig*, and *bark*.

Procedure

[Note: This lesson can be repeated in different locations and at different times of the year.]

Getting Started

Curiosity and Prior Knowledge: To spark children's interest, let the children know that they will be going on a nature walk and that during the walk they will collect things from nature to make "nature bracelets." To help your children better understand what they'll be making, make your own simple nature bracelet to show as a model. Before heading outside, have children discuss the things they expect to find for their nature bracelets. Ask, "What kinds of items do you think you'll find to stick on your bracelets?" As children share their ideas, guide children toward natural objects rather than human-made ones. Make a list of these objects on chart paper or poster board; be sure this list is visible to children and that words are accompanied by picture support. Title this list "Things we think we'll find in nature" and put it aside to compare at the end of the lesson. Then prepare for your nature walk; wrap masking tape around children's wrists with the sticky side of the tape facing outward so that during the walk children can collect and adhere items to their bracelets. [Note: Have children create two bracelets: one bracelet to be used during the science lesson and one to keep and take home.]

Investigating 1

Observing: Review safety precautions with children and then take them on the nature walk. As children explore and collect natural items for their bracelets, engage children in conversations about the items they have found. Ask questions such as "What different colors have you found? Did you find anything that was part of a tree? What part of the tree did it come from? What is the best thing you collected? Why do you like that item best?" These conversations are important for developing oral language as well as for increasing interest in nature and the items children have found.

After the walk, have children select one of their bracelets for investigating and one to take home. Explain that children will carefully take apart and observe the individual parts of one of their bracelets. The bracelet that each child chooses to keep should be put away for safekeeping, but the one chosen for investigating should be removed and placed in front of the child. Arrange children in small, cooperative groups and give each group a tray. Ask the children to take each item from their bracelets and place it into the tray. Provide magnifiers and prompt children to make observations of each individual item. As you did during the nature walk, ask individual children questions to encourage them to describe the items they've collected.

Comparing and Sorting: After children have had an opportunity to make careful observations of their items, ask children to sort the items by putting items that are similar into groups. Let children sort as they wish and ask them to describe how they sorted. Once this initial child-directed sorting is complete, give children categories and ask them to sort again. You can use just two categories ("items from plants" and "items not from plants") or several categories (e.g., "rocks and soil," "plants," "human-made," etc.) as

Nature bracelets

appropriate for your children. [Note: You may need to alter these sample categories based on the specific items that your children have collected.] As children sort, use questioning to encourage children to describe, and even name, the items and to explain why they have sorted the items as they have.

Investigating 2

Observing and Sorting: Next, focus on the "plant" items, placing all other items aside. Have children share the items in the plant category and as they do, ask children questions such as "What is this item? How do you know it is part of a plant? How do you think this helps a plant?" Emphasize children's explanations and validate their thinking, keeping in mind that for young learners, an enthusiasm for science is more important than the "right" answer. Ask children to sort the plant items into two categories ("items from flowers" and "items not from flowers"). As in *Investigating 1*, while children sort, prompt them to describe each item, to compare different items to notice similarities, and to explain why they have placed an item into a specific category. Ask questions such as, "Can you show me two items that are similar? How are they alike? Can you show me something that is different? How is it different?" In their "items from flowers" category, children will have several different flowers as well as different flower parts (e.g., individual petals).

Making Sense

Communicating Findings: Start a new list titled, "Things that we found in nature" and on this list write the word *flowers*. Have children help spell, read, or identify letters and letter sounds in the word *flower*, as appropriate for their language development. Then ask children, "Does anyone have a whole flower?" and have children tape a few examples of flowers to the list. Children will also have individual petals and maybe even other flower parts. For these, ask children, "Is that part of a flower?" followed by "What do you think it helps the flower do?" Ask for additional examples of this flower part and tape them to the list. If appropriate, you may introduce some flower vocabulary at this point (e.g., "These colorful parts of the flower are called *petals*"), adding labels to your list. Review your list and have children identify whole flowers. Then ask them to locate flower parts, asking, "Flowers have different parts; do we have any parts of flowers on our board?" Following your children's interest, repeat the *Investigating 2* and *Making Sense* procedures for other plant parts they may have collected such as twigs, leaves, bark, etc.

Curiosity: At the completion of this lesson, you and your children can review the "Things we think we'll find in nature" and the "Things we found in nature" lists. Engage children in a conversation about items they anticipated finding but did not find; for example: "We thought we'd find feathers on our nature walk, but we didn't find any. I wonder why? Can you think of an explanation? Where do you think we'd be more likely to find feathers? Why?" Use this conversation to help spark children's curiosity about nature and to promote their critical-thinking skills.

What's Next?

Extension Activity

If you complete this lesson in the spring, children will likely select several flowers and flower parts for their nature bracelets. Capitalize on this to help children to explore the structure of flowers. Provide each child with one large flower; the parts on these larger flowers will be easier for children to observe and manipulate. Have children "dissect" the flowers. [Note: If *dissect* is a new word for your class, take time to support children's understanding of it. Show the word, have children identify different letters and letter sounds, and say *dissect* together. Let children know that to dissect means to take something apart and examine its individuals parts. Use the new term throughout the lesson and encourage your children to do the same.] Flowers generally have four distinct structures: sepals, petals, stamen, and pistil, but learning the names and functions of each of these parts isn't of primary importance. The learning objective here is for children to understand that flowers have different parts. Provide children with a piece of paper with a large circle divided into four sections drawn on it. Help them to title this page "Flower Parts" and encourage them to place, and later glue, similar parts into the same section of the circle. Afterward, you can show children flowers or pictures of flowers (ideally including the flowers that children encountered during their nature walk). In these flowers, older children can identify the flower parts they discovered through dissection, and younger children can count the 1, 2, 3, 4 parts of each flower. [Note: Not all flowers have all four parts. Be selective in your selection of flowers so that children have multiple experiences with the common structure of flowers before encountering the very many exceptions later in their education.]

Integration to Other Content Areas

Reading Connections

This lesson has many opportunities for children to develop their emergent literacy skills to learn about flowers, as well as nature in general. Fill the shelves of your classroom library with both fiction and nonfiction books about flowers and nature. For flowers, consider books that introduce children to a variety of flowers, such as *The Flower Alphabet Book* by Jerry Pallotta (1989), *Planting a Rainbow* by Lois Ehlert (2003), and *Flower Garden* by Eve Bunting (2000), and share stories about children planting and tending to flower gardens, such as *Lola Plants a Garden* by Anna McQuinn (2017), *My Garden* by Kevin Henkes (2010), *Wanda's Roses* by Pat Brisson (2000), and *In the Garden* by Elizabeth Spurr (2012). For nature, consider books such as *I Took a Walk* (1998) and *On the Way to the Beach* (2003) both by Henry Cole and *In the Tall, Tall Grass* by Denise Fleming (1995). (See *Scavenger Hunt*, p. 177, for even more books about nature to share with your children.)

Writing Connections

Children can develop their fine motor skills by tracing flowers you have drawn. Draw flowers with long, squiggly stems to make tracing more challenging. Have your children complete this task with different-sized flower drawings and with different media (e.g., crayons, chalk, pencils, paint, and so on.). Cut 26 small flower shapes out of construction paper and write a different letter of the alphabet on each. Children can identify the letters and sequence your flowers in alphabetical order or arrange them to spell words. Provide children with the words for the items they collected on their nature bracelets, such as leaf, twig, stick, petal, acorn, bark, etc., and encourage children to spell out each word using your flower letters.

Math Connections

Flowers present several opportunities for children to count. Children can count the petals of individual flowers and sort the flowers into five groups—flowers with three petals, flowers with four petals, flowers with five petals, flowers with six petals, and flowers with more than six petals. Create a counting game with dice, flowers (artificial flowers will do), and several small vases. Have the children roll a die, count the number on the die, and then count out that many flowers to put into a vase. Repeat the process until all flowers are used or the vases become full.

Other Connections

Child's Life Connection

Children can create a weeklong nature journal. Each day, give children a piece of paper with the day of the week written at the top. Children can trace the day of the week and then collect one item from nature (leaf, flower, piece of bark, etc.) to glue or tape to the center of the paper. Children can tell an adult the name of the item and what is important about it (e.g., why she or he selected it, where it was found, etc.) to be written beneath the item at the bottom of the paper. Fasten the sheets together to create a nature book. Help children to read their book by saying with them, "On Monday, I collected a ___" and pointing to the words that represent the day of the week and the name of the item as you read. Read aloud the first few days together and then encourage the child to read aloud the last few days on his or her own.

Center Connections

At the **art center,** children can make paper flowers. There are countless different flower-making art projects that are appropriate for young learners. Try making coffee filter flowers. Children can use watercolors to paint the filters. Once dry, children can fold the filter and twist a green pipe cleaner stem around the center of the filter. At the **sensory table**, put out several different flowers for children to explore. Encourage children to use multiple senses (e.g., sight, smell, touch) as they observe the flowers and to notice the different textures, colors, and shapes of flowers. Providing magnifiers will help children to look closely and see even more detail. For **dramatic play**, children can pretend to be florists. Provide vases, a variety of silk flowers and plant parts, and ribbons and bows. Children can assemble bouquets and other flower arrangements and present them to each other. They can also tell each other about the important events that the flowers are for.

Family Activities

This week at school, your child has been exploring nature. You can help your child to become more aware of and interested in nature by finding time to explore as a family. Although a family camping trip to a wilderness area would be great, nature explorations can be far more accessible than that. Simply go to a local park, walk around your neighborhood, or visit any area where there are different plants to explore. Let your child show you how to make nature bracelets (i.e., a piece of masking tape wrapped around your wrist with the sticky side facing out). Everyone in your family, young and old, can find natural items such as leaves and petals and twigs to stick to your bracelets. Then everyone can share his or her beautiful creation. Have your child help you share, asking him or her to describe each item and where it came from. Enjoy noticing the beauty in nature together. Attached is a map of parks and other natural areas nearby.

Actividades Familiares

Esta semana en la escuela, su hijo ha estado explorando la naturaleza. Puede ayudar a su hijo le a ser más consciente de o interesarse más en ella la naturaleza encontrando dejando tiempo para explorar como familia. Mientras que un viaje de campamento familiar a un área silvestre sería genial, explorar la naturaleza puede ser mucho más accesible que eso. Simplemente ir a un parque local, caminar por su vecindario o visitar cualquier área donde hay diferentes plantas para explorar. Deje que su hijo le muestre cómo hacer pulseras naturales (es decir, un trozo de cinta adhesiva envuelta alrededor de su muñeca con el lado adhesivo hacia fuera). Todos en su familia, jóvenes y viejos,mayores pueden encontrar elementos naturales como hojas, pétalos y ramas que se adhieran a sus pulseras. Luego, todos puede compartir su hermosa creación. Haga que su hijo le ayude a compartir, pidiéndole que describa cada elemento ypieza de dónde viene. Disfruten de apreciar la belleza de la naturaleza juntos. Aquí le presentamos un mapa de parques y otras áreas naturales cercanas.

[Note: Attach a map indicating parks in your local area here.]

Assessment—What to Look For

- **Can children describe objects from their nature-based observations?**
  (using senses to gather information, identifying properties, using complex patterns of speech)
- **Can children sort objects based on observable characteristics?**
  (identifying properties, comparing properties, classifying)
- **Can children explain their reasoning?**
  (using complex patterns of speech, constructing explanations, explaining based on evidence)

Standards

Head Start Early Learning Outcomes Framework
P-SCI 2. Child engages in scientific talk. "Uses scientific content words when investigating and describing observable phenomena, such as parts of a plant, animal, or object."
Next Generation Science Standards
Science and Engineering Practice: Planning and carrying out investigations. "Make observations (firsthand or from media) and/or measurements to collect data that can be used to make comparisons. Use observations (firsthand or from media) to describe patterns and/or relationships in the natural and designed world(s) in order to answer scientific questions and solve problems."
Common Core State Standards for Mathematics
1.MD.C.4. Represent and interpret data. "Organize, represent, and interpret data with up to three categories; ask and answer questions about the total number of data points, how many in each category, and how many more or less are in one category than in another."
Common Core State Standards for English Language Arts
L.K.5.A. Vocabulary acquisition and use. "Sort common objects into categories (e.g., shapes, foods) to gain a sense of the concepts the categories represent."

Finding Weeds

with Myra Pasquier

Lesson: Mapping where weeds grow

Learning Goals: Children will investigate the school grounds to find different weeds and to identify the different places weeds can grow.

Materials: Magnifiers, pictures and samples of dandelions and other weeds found at your school, maps of your schoolyard, and drawing and writing materials

Safety: This lesson will involve being outside. As with all field trips, scout out the area before you take children there; make sure the area is safe and free from hazards. Be mindful of roots, obstructions, and other hazards along your path. Be aware that many weeds have bristles, spines, and thorns that can poke, scratch, or irritate children's skin. Also, dandelion sap can cause skin irritations; children should wash their hands thoroughly after handling plants. Be aware of children in your class who may have allergies (pollen, bees, etc.) and plan accordingly. Be careful to stay away from poisonous plants. It will be helpful to have additional adults to help supervise and interact with children while you are all outside.

Teacher Content Background: Generally speaking, there are two life strategies for plants: (1) find the ideal location, grow slowly, and be long-lived and (2) find any place that will work, grow quickly, and be short-lived. Weeds typically use the second strategy and are very successful at growing and thriving in different environments because of several adaptations, including producing many small seeds; dispersing seeds over relatively great distances; and growing and maturing rapidly. Although weeds can be considered a nuisance in gardens, they are perfect examples of resiliency and adaptation in nature.

One of our most well-known and beautiful weeds, the dandelion, flowers throughout the late spring, summer, and well into autumn. It's yellow "flowers" are actually a large grouping of dozens of individual flowers, each capable of becoming its own individual fruit. These familiar fruits each have fluffy, feathery bristles that act as sails, helping the seed to float in the breeze. Wind can disperse these fruits as far as 5 miles, explaining why dandelions reappear in your yard year after year. If you decide to eliminate the dandelions that pop up, be sure to pull up the plant's taproot. These deep roots can grow to be more than a foot long and, when the aboveground parts of the plant are removed, can sprout new leaves.

Science terms that may be helpful for teachers to know during this lesson include *observe*, *map*, and *weed*.

Procedure

[Note: In this lesson, you and your children will be taking a class trip around your school to map where weeds grow. To prepare for this lesson, scout out your schoolyard for weeds and plan a route that features weeds growing in many different areas (e.g., in planter beds, in grassy fields, through cracks in the pavement, etc.). Take photographs of the different weeds you find and learn the names and key features of the most common ones.

Your children will need some assistance as they make a map of their schoolyard to show where weeds are found. Younger children will need mostly completed maps but can trace the route you take on your trip and mark the locations of weeds you find together. Encourage and support older children to create their own maps. Children's map-making can be completed before or embedded within this lesson.]

Getting Started

Introduction: Collect and bring to class a few samples of dandelions; be sure your samples include leaves, flowers, and "puff balls" of seeds. Show these samples to children and ask children if they've ever seen this type of plant before. Have a conversation with children as they share what they know about and their experiences with this plant. Once children have had an opportunity to share, direct the conversation toward where dandelions grow, asking children if they have any ideas or questions about where dandelions grow. To extend the conversation and children's thinking about dandelions, ask additional questions, such as "Where have you seen dandelions? Where do dandelions grow? Can we find dandelions here at school? Where?" Using words and picture support, record children's ideas in a visible place.

Initial Explanation: The idea that dandelions are weeds may arise during your introductory conversation. If so, follow up with this idea at that time. If "weeds" do not come up during your conversation, prompt children by asking, "Some people call dandelions and some other plants weeds. Does anyone know what a weed is?" Keep track of children's ideas, and at the end of this discussion summarize and record their ideas as the class's initial definition of weeds.

Investigating

Observing: Let the children know that they will be going on a "weed walk"—searching for weeds around the schoolyard. Share samples (collected away from school) and photographs of the four or five most common weeds at your school. [Note: Some weeds have spines; thorns; or stiff, irritating hairs. Do not include these in your weed explorations with children.] Write the name of the weed on each picture for children to read. After this introduction, encourage children to describe each weed. Record their descriptions. After each weed has been observed individually, invite children to compare the different weeds and to discuss how they can tell them apart.

Before starting on your "weed walk," children will need to make or be oriented to maps of the schoolyard. Share a sample map and show children a few key features (e.g., your classroom, the office, the parking lot). Demonstrate to children how to locate additional places on the map. "I know the flagpole is between the parking lot and the office. If I wanted to show on my map where the flagpole is, I would point to right here." And then challenge children to work with a partner to discuss and find the location of other places on their maps (the playground, the cafeteria, etc.). With older children, provide graph paper and time for them to create their own maps of the school. Once all children have and understand their maps, have children locate their starting point and begin your walk. During the "weed walk," ask children to look for weeds and stop often to give children a chance to look closely. Each time a weed is found, have children mark on their map where the weed was found and trace their path to the weed.

Making Sense

Generating Explanations: After your "weed walk" is complete, have children analyze their maps. Encourage children to discuss where weeds are found and to consider how weeds got to these many different places. Ask children, "How do you think this plant got here? Do you think the gardener planted this here?" Challenge children's thinking with questions such as "Can all plants get to and grow in these different places? How come there aren't oak trees growing in cracks in the sidewalk?" The goal here is to provide opportunities for children to come to understand that weeds are plants that are able to get to and grow in many different places. Refrain from giving children this textbook definition. In all that we do with young children, children's thinking—their ideas and their wonderings—should be the focus. Revisit the initial definition of weeds generated during *Getting Started* section and ask children if there is any information that they would like to add (or remove) from their definition. If necessary, prompt children to consider the many different places where weeds can grow. Complete and review together your class definition of weeds.

What's Next?

Extension Activity

Dandelions make for great plants to explore, in part because of their large taproots. Collect several samples, roots and all. [Note: Dandelion roots can extend for more than a foot into the soil; use a small shovel or other gardening tool to ensure that you collect the entire plant.] In the classroom, children can make observations of the collected weeds. Remind children of the basic parts of a plant (roots, leaves, flowers, etc.) and as children make observations, they can draw (or take and print photographs of) the dandelion and label each of the different plant parts. [Note: *Where Vegetables Come From*, p. 88, also engages children in an investigation of different plant parts.] Encourage children to measure (using standard or nonstandard units) the length of the different parts and help them to represent these lengths in their drawings and photographs. Children can compare their measurements to see who has the longest root, the tallest flower, the longest leaf, and so on.

Integration to Other Content Areas

Reading Connection

To further emerging literacy, provide both fiction and nonfiction books for children to explore. In selecting books, consider one of our most common weeds—the noble, resilient, and beautiful dandelion. Engaging storybooks about dandelions include *The Dandelion's Tale* by Kevin Sheehan (2014), *The Dandelion Seed* by Joseph Anthony (1997), and *Dandelions* by Katrina McKelvey and (2015). While nonfiction books, such as *Dandelions:*

Stars in the Grass by Mia Posada (2000), *A Dandelion's Life* by John Himmelman (1999), and *Dandelions* by Kathleen V. Kudlinski (1999), will help you and your children learn more about these amazing plants. Finally, a defining characteristic of weeds is their ability to thrive under conditions that many other plants would not. Children can learn about the resiliency of weeds, along with their many other fine attributes, by reading books such as *Dandelion Adventures* by Patricia Kite (1998) and *Weeds Find a Way* by Cindy Jenson-Elliott (2014). As you read these books, fiction or nonfiction, encourage children to talk about what they know about weeds and to share any new questions they have or insights they learned from the book.

Writing Connections

Encouraging children to write or draw their observations supports their writing development and their understanding that print has meaning. In this lesson, children engaged in meaningful writing by drawing and labeling maps with landmarks and the different locations of weeds found during their walk. You can extend this learning by sharing various maps with children. Together, you can identify the features of maps. Older children can then add a legend, scale, and title to their "weed walk" maps; younger children can collaborate with you to create a whole-class map with these features. Also, "weed walks" provide great opportunities for exploring the letter *W* and for exploring alliterations. Have children name all the *W* words or all of the two-word alliterations (real or inventive) they can think of. Use words and picture support as you record their ideas.

Math Connections

Maps are great tools for promoting mathematical thinking. Provide simple maps of familiar areas (e.g., the layout of your school or local library, a simple street map of the neighborhood around your school, etc.) and challenge children to find a route from one point to another and to measure the distance using standard or nonstandard (e.g., paper clips, counting blocks, etc.) units. Children can measure and compare multiple routes to find the shortest distance between two points. Let children decide where on the map they want to set as their destination(s). Children can share and describe the routes they plan. As developmentally appropriate, help children to develop directionality using ideas such as right and left; forward and backward; and north, south, east, and west.

Other Connections

Child's Life Connection

To better appreciate the presence of weeds in their daily lives, children can create additional "weed maps" of their schoolyard or of other places—perhaps the route between their home and school or around their apartment complex. The conspicuous dandelion may be the best weed for children to map. Let children decide the areas they want to map. Children can draw maps to the best of their ability and describe to you what their map shows. Give children other dandelion-mapping challenges such as counting the dandelion flower heads in a field or the number of houses in their neighborhood that have dandelions in their lawns, and ask them to draw maps that represent these data. In addition to finding weeds in their daily lives, children can explore maps. Gather maps of local amusement parks, zoos, parks, museums, and so on, for children to "read" and discuss.

Center Connections

In the **art center**, children can use dandelions as paintbrushes to paint a picture of a dandelion. Invite children to disassemble a dandelion (i.e., separate roots, leaves, flower stalks, and flowers) and to then dip each of the parts in paint and press them onto white construction paper. [Note: This art center activity is messy.] The flowers of weeds may be the most readily accessible of all flowers. Have children help you find and pick these flowers for your **sensory table**. There, children can use magnifiers to explore the flowers. Although most are smaller than the flowers you'll find at a florist, they come in a wide array of colors and shapes. Encourage your children to compare flowers to find similarities and differences. And help them to see and appreciate these small and beautiful wonders of nature. For **dramatic play**, children can hide treasures and make treasure maps for each other to read and follow. You can initiate this by hiding a treasure in your classroom and creating a (sample) map for children to follow. Provide drawing materials, including paper with an outline of your class layout, and treasures for children to hide.

Family Activities

At school, your child and her or his classmates have been looking for and learning about dandelions and other weeds. You can participate in your child's learning by taking a "weed walk" together. Weeds are plants that are able to get to and grow in many different places, including places they may not be wanted. On your walk, you and your child can search for weeds in different places. When you find a weed, ask your child how she or he thinks the weed got there and where else she or he thinks this kind of weed might be found. Listen to your child's ideas and share some ideas of your own. If you are fortunate enough to find several dandelions, you and your child can pick the flowers to make a beautiful bouquet and make wishes together as you blow dandelion seeds. Watch or follow after the seeds as they float away and ask your child how far she or he thinks the seeds might go and what might happen to them once they land. After your weed walk, you and your child can search for books about dandelions at your local library; read them together to learn more about these plants.

Actividades Familiares

En la escuela, su hijo y sus compañeros han estado buscando y aprendiendo sobre las plantas delos dientes de león y otras hierbas. Usted pPuede participar en el aprendizaje de su hijo llevándolo a dar un "paseo de malas hierbas malas" juntos. Las malas hierbas malas son plantas capaces de crecer en muchos lugares diferentes, incluyendo aquellos donde pueden no ser bienvenidas. Durante el paseo, usted y su hijo pueden buscar las hierbas malas en distintos lugares. Cuando encuentre una mala hierba mala, pregúntele a su hijo cómo piensa que la mala hierba mala llegó allí y donde más cree que se puede encontrar este tipo de malas hierbas. Escuche las ideas de su hijo y comparta algunas ideas propias. Si tiene la suerte de encontrar varias plantas de os dientes de león, usted y su hijo pueden elegir recojer las flores para hacer un ramo hermoso y pedir deseos juntos al soplar las semillas de diente de león. Vea o siga las semillas, ya quecuando están volando flotan lejos y pregúntele a su hijo hasta que punto cree que pueden llegar las semillas y lo que podría pasar con ellas una vez que caen se depositan en la tierra. Después de dar un paseo de malas hierbas malas, usted y su hijo pueden buscar libros sobre las plantas de os dientes de león en su biblioteca local; léanlos juntos para aprender más sobre estas plantas.

Assessment—What to Look For

- **Can children describe objects from their nature-based observations?**
  (using senses to gather information, identifying properties, using complex patterns of speech)
- **Can children sort objects based on observable characteristics?**
  (identifying properties, comparing properties, classifying)
- **Can children explain their reasoning?**
  (using complex patterns of speech, constructing explanations, explaining based on evidence)

Standards

Head Start Early Learning Outcomes Framework
P-SCI 6. Child analyzes results, draws conclusions, and communicates results. "Communicates results, solutions, and conclusions through a variety of methods such as telling an adult that plants need water to grow or putting dots on a map that show the number of children from each country."
Next Generation Science Standards
Science and Engineering Practice: Planning and carrying out investigations. "Make observations (firsthand or from media) and/or measurements to collect data that can be used to make comparisons. Use observations (firsthand or from media) to describe patterns and/or relationships in the natural and designed world(s) in order to answer scientific questions and solve problems."
Common Core State Standards for Mathematics
K.G.A.1. Identify and describe shapes. "Describe objects in the environment using names of shapes, and describe the relative positions of these objects using terms such as above, below, beside, in front of, behind, and next to."
Common Core State Standards for English Language Arts
W.K.3. Text types and purposes. "Use a combination of drawing, dictating, and writing to narrate a single event or several loosely linked events, tell about the events in the order in which they occurred, and provide a reaction to what happened."

Branch Puzzles

with Nicole Hawke

Lesson: Putting pieces of a branch together to make a whole branch

Learning Objectives: Children will use their observation skills to identify what tree branches look like. They will then use the knowledge they gained from their observations to put pieces of a tree branch back together.

Materials: Several sets of branch puzzles, a camera, drawing and writing materials

Safety: Prepare the puzzles outside of class time and away from children. Wear safety goggles or safety glasses and be careful as you cut branches into pieces for the puzzles. This lesson will involve being outside. As with all field trips, scout out the area before you take children there; make sure the area is safe and free from hazards. Be aware of children in your class who may have allergies (pollen, bees, etc.) and plan accordingly. Be careful to stay away from poisonous plants. Also, be mindful of roots, obstructions, and other hazards along your path outside. It will be helpful to have additional adults to help supervise and interact with children while you are all outside.

Teacher Content Background: Trees are some of the oldest and largest living organisms on Earth (a giant sequoia, named General Sherman is estimated to weigh more than 2,000 tons and a bristlecone pine, named Methuselah, is nearly 5,000 years old). The massive growth of trees is accomplished through photosynthesis—the process by which the carbon of carbon dioxide (CO_2) in the air is captured and used to create the molecules that make up the tree. (Yes, the vast bulk of trees, and therefore wood, was once CO_2 in the air!) Trees can grow to amazing heights; some of the world's tallest trees are more than 300 feet tall—taller than a house, a water tower, and even the Statue of Liberty! In addition to growing taller, each year a tree will produce new growth in the trunk outside of its previous growth, increasing the diameter of the tree and contributing to a ring-like pattern that scientists use to determine the age of a tree. Outside of this new growth of wood (which serves to transport water and minerals between the roots and leaves), trees grow bark. The innermost layer of bark works to transport "food," made through photosynthesis, from the leaves to the rest of the tree. However, the vast majority of bark is made up of dead cells, serving to protect the tree from disease and insects.

Science terms that may be helpful for teachers to know during this lesson include *observe, compare, tree, trunk, branch,* and *bark.*

Procedure

[Note: To prepare for this lesson, you'll have to create branch puzzles. Cut multiple, 2-foot lengths from fallen tree branches and remove all leaves and side branches; the thicker end of each should be 2 to 3 inches in diameter. Cut each length of branch into three or more pieces. The more pieces there are, the harder the puzzle will be. (Having a few challenging puzzles prepared may be useful if you have children who find the three-piece puzzle too easy.) Use some straight cuts across the branch and some angled cuts. Be sure to bag or label each set of pieces immediately so they don't get mixed up. Throughout this lesson, use branches that have fallen or have been pruned; do not harm trees for this lesson.]

Getting Started

Prior Knowledge: To begin this investigation, ask children to think about trees and to share with the class their personal knowledge about trees, as well as any questions they have about trees. Invite children to describe a tree to you. As they are discussing their mental image of a tree, record the different parts or other ideas they are mentioning (leaves, twigs, trunk, flowers, green leaves, brown sticks, etc.) on large piece of paper, on a whiteboard, or on a drawing of a tree.

Child Questions: Take your children for a "wonder walk" around the school grounds or at a nearby park to an area that has numerous trees. [Note: Before leaving the classroom, briefly discuss with the children any safety concerns specific to the walk you will be taking.] Invite children to stop and look closely at the trees. Encourage them to think of questions they might have about trees. Start by sharing some of your own wonderings, such as "I wonder which part of the tree is the oldest" or "I wonder where the largest leaf is on this tree." Don't concern yourself or the children with right answers; curiosity, questioning, and exploring are far more important for our young learners. After children have had an opportunity to wonder about various aspects of the tree, direct their interest toward the tree's branches. Ask prompting questions such as "Which part of the tree is the biggest? Can your arms reach all the way around the trunk? Can your hands reach around a branch? Are the tree's branches all the same? How do the branches on the tree change from where they connect to the trunk to where they end?" When back in the classroom, invite them to remember their questions and observations of the tree. Record these ideas, adding them to your initial notes to be referred to later in the lesson.

Investigating

Observing: Draw children's attention to the ideas about branches generated during the wonder walk. Then provide each pair of children with a bag of branch pieces. Encourage the children to remove the pieces and look at each piece carefully. After these initial observations, tell children that these are "branch puzzles." Describe how the puzzles were made and explain that just like the puzzles they play with, the pieces of a tree branch puzzles need to be put back together. Prompt children to see if they can piece the branch back together again. [Note: You may have to model putting a branch puzzle together before you let children attempt their puzzles.] As you circulate among the groups, you can scaffold groups that may be struggling by asking them to think back to the branch observations the class made, asking, "How did braches change from beginning to end?" For pairs that get the puzzle quickly, challenge them with a puzzle that has more pieces or a puzzle made up of pieces from more than one branch to see if they can sort out the pieces to make two branches.

Documenting: After they piece their puzzles back together, have children draw a model of their branch puzzle, tracing the connected pieces on paper. These models could be shared with other partner pairs as children explain how they figured out how the puzzle fit together. Encourage children to examine how the size of the branch varies from one end to the other, and ask that they make this observation of other children's puzzles as well.

Making Sense

Identifying Patterns: Once children have examined and compared different branch puzzles, have a conversation about their findings. Ask children how they were able to solve their puzzle and how they figured out the sequence of the pieces. Show children a puzzle that you have completed incorrectly with the thinnest end connected to the widest end. Have children describe why it's wrong. (If children do not find fault with your puzzle by looking at it, invite them to feel your puzzle. They should notice that the branch pieces connect evenly, except at your incorrect connection; if not, point this out. Have a child or two reconnect your puzzle correctly and invite children to look at and feel the puzzle again to notice the difference.) As children describe their correct and your incorrect puzzle, draw their attention to the ends of the branch. Help children to describe one end as thin and the opposite end as thick. Children can present their model to the class, showing that one end of their branch is thin and that the other end is thick.

Putting together a branch puzzle

Application: To observe additional examples of this pattern, take the children on another walk to observe trees. (Perhaps they can observe their adopted tree—see *Adopt a Tree*, p. 164.) Give your children time to observe any features of the tree that they may notice. As they are observing, refer to your notes from the *Getting Started* section to link back to some of the things the children told you that they knew about trees. Eventually, direct children toward the branches of the tree; ask them what they notice about the size of the branch. If possible, find a low-hanging branch and have children grab the branch toward its tip and then grab the branch again near the tree's trunk. Have children describe the difference and help them make the connection to the branches used in their puzzles. Children can also take and analyze photographs of the tree branches. Children should see that branches are thicker near the tree's trunk; once you're back in the classroom, have children amend their branch puzzle models to show which end would be closest to the trunk.

What's Next?

Extension Activity

Trees are the largest, most obvious living things that children experience in their daily lives and are therefore great opportunities for science explorations. During the spring or summer, ask children to look carefully at trees to search for animals such as birds, squirrels, insects, and spiders and for evidence that animals have used the tree. Encourage children to look for birds' nests, spiderwebs, and leaves or fruits that have been nibbled. Ask them to think about and describe different ways that trees can help animals. After searching for animals in trees, share books such as *The Busy Tree* by Jennifer Ward (2009), *Who Lives in Trees?* by Trish Holand (2010), *Little Acorn Grows Up* by Edward Gibbs (2013), or *One Small Place in a Tree* by Barbara Brenner (2004) to expand children's ideas about the ways different animals rely on trees.

Integration to Other Content Areas

Reading Connections

Through multiple opportunities to participate in classroom discussions and share their observations with classmates, children are fostering their oral language development—an important precursor for reading development. Children's reading interest and ability can be further developed by incorporating several different books about trees in your teaching. Informational books about trees and their parts, such as *A Tree Is a Plant* by Clyde Robert Bulla (2016) and *Tell Me, Tree: All About Trees for Kids* by Gail Gibbons (2002), will provide children with opportunities to expand their understanding of and curiosity about trees while experiencing the features of informational text. Sharing books such as *We Planted a Tree* by Diane Muldrow (2016), *The Tree Lady* by H. Joseph Hopkins (2013), *The Tree* by Dana Lyons (2002), and *Wangari's Tree of Peace: A True Story From Africa* by Jeanette Winter (2008) will help children to understand that people have an important role in protecting trees and other living things. Tree-centered stories, such as *Stuck* and *The Great Paper Caper* both by Oliver Jeffers (2011, 2009), *A Tree for Emmy* by Mary Ann Rodman (2009), and *Pablo's Tree* by Pat Mora (1994), can help children to see that science topics can be used in creative ways.

Writing Connections

Sequencing is an important skill for writing. Writing is therefore similar to the branch puzzle activity in that letters and words need to be arranged in the appropriate order. Provide children with "word puzzles" (e.g., a bag containing the letters, *A-B-C-H-N-R*) and let them sequence the letters in alphabetical order and arrange the letters into words (*BRANCH*, but also *RAN*,

CAB, etc.). Provide several different tree-related puzzles and a list of "answers" (e.g., *Tree, Leaf, Branch, Bark, Wood*). Children can then arrange the letters of each puzzle to match one of the tree-related "answers." For younger children, you can have similar puzzles with the letters of the child's name. Each child can complete his or her name puzzle by arranging the letters of his or her name in the correct order. In creative writing, children can make up a story about being a tree or about an animal that makes its home in a tree. After children have written or drawn their stories, provide children with an opportunity to "read" their stories to help reinforce children's understanding that print has meaning.

Math Connections

Trees are great sources of math explorations with opportunities for children to count, to identify shapes, and especially to measure. Children can count the number of trees in an area or the last few leaves hanging onto a deciduous tree in December, and they can search for trees and tree shadows that resemble different shapes (triangle, oval, etc.). Using both standard and nonstandard units, children can measure different parts of trees (branches, leaves, trunk width, etc.). Children can also measure the length of a tree's shadow and compare shadows as an indication of trees' relative height. Children can begin to compare size when measuring different parts of the tree, making important discoveries such as "Some leaves are big and some leaves are small" and "The trunk is bigger around down at the bottom of the tree."

Other Connections

Child's Life Connections

Set out a small collection of a few different leaves, tree fruits or seeds (acorns, pinecones, etc.), branches, or bark pieces. Have the children carefully observe and touch the items. Invite children to add to this collection. Children can collect small parts of trees from their homes or neighborhoods. As children bring different plant parts in, encourage them to describe what they've brought in and the tree that it came from. Use question sequences to encourage children to generate descriptions that require more complex thinking and language use (e.g., "Tell me about the tree you observed. How tall was the tree that your leaf came from? Was the tree taller than you? Taller than a house?" and "Can you describe the leaves on the tree you collected from? How big were the leaves? Were they bigger or smaller than your hand? What shape were the leaves? Could you draw a picture of the leaf?").

Center Connections

At the **art center**, provide children with a variety of different types of brownish paper. Have children tear the paper into pieces and glue them into the shape of a trunk and tree branch. If children require additional support, draw an outline of a tree on paper for children to glue the brown paper to. Once the "trunk" and "branches" are dry, provide children with different types of greenish paper and ask children to tear and glue "leaves" to their tree. (Children can work on their own individual trees or use butcher paper to create a large class tree.) At the **sensory table**, add branches from several different types of trees with different types of bark. Allow children to observe the branches, focusing on how the branches look and feel. As children explore the branches, encourage them to use language to describe texture, color, size, and so on. Ask questions such as "How does the branch feel? What different colors do you see?" In your **dramatic play** area, put up a large, recently cut tree branch with many side branches and leaves. Allow your children to "populate" the tree. Set out stuffed animals and toy birds, insects, and spiders. Encourage children to describe how the different animals might use the tree. Provide materials for children to build birds' nests and spiderwebs. Ask children what other animals or materials they need to place in their tree. Providing hiking clothes and hats, bandanas, and boots will allow children to take on the role of hiker, camper, or park ranger.

Family Activities

In school, your child has been learning about trees. You can further your child's interest in trees by planting and caring for a tree with your child. With your child, decide upon a tree that is best for your home. You could choose a tree that can be grown in a pot on the patio, one that could live indoors, or one that could be planted in the yard. Start with a small, young tree or even try growing your tree from a seed. Ask your child what the tree will need to grow. As much as possible, let his or her ideas direct the selection, planting, and care of your tree. (If planting a new tree isn't possible, select an existing tree from your yard or neighborhood.) Make caring for your tree a special time that you share with your child. As your tree grows and changes with the seasons, talk with your child about the changes that he or she notices and encourage him or her to wonder how these changes are happening. The books, Red Leaf, Yellow Leaf by Lois Ehlert, Pablo's Tree by Pat Mora, and We Planted a Tree by Diane Muldrow tell the story of children planting and enjoying trees with their families; reading these or similar books about trees with your child can prompt further conversations about the tree that you planted together.

Actividades Familiares

En la escuela, su hijo ha estado aprendiendo sobre los árboles. Puede fomentar el interés del niño en ellossobre los arboles plantando y cuidando de un árbol con su hijo. Con su hijo, decida sobre qué árbol es mejor para su hogar. Puede optar por un árbol que se pueda cultivar en una maceta en el patio, o uno que pueda vivir en el al interior, o uno que se pueda plantar en el patio. Comience con un árbol pequeño, joven con un árbol pequeño o incluso intente hacer crecer su árbol a partir de una semilla. Pregúntele a su hijo lo que necesita el árbol para crecer. Tanto como sea posible, deje que las ideas de su hijosus ideas guíen la elección, la plantación y el cuidado de su árbol. (Si no es posible plantar un nuevo árbol, seleccione un árbol existente de su jardín o en el vecindario). Haga que el cuidado de su árbol sea un momento especial que comparta con su hijo. A medida que el árbol crece y cambia con las estaciones del año, hable con su hijo sobre los cambios que nota y anímelo a preguntarse cómo estos cambios están sucediendo. Los libros Hoja roja, hoja amarilla de Lois Ehlert, El árbol de Pablo por Pat Mora y Plantamos un árbol de Diane Muldrow cuentan la historia de hijos niños que plantan y disfrutan de los árboles con sus familias; la lectura de estos libros u otros similares sobre los árboles con su hijo puede fomentar más conversaciones sobre el árbol que plantaron juntos.

Assessment—What to Look For

- **Can children ask questions about trees?**
  (asking questions, identifying properties, making inferences)
- **Can children describe trees based on their observations?**
  (using senses to gather information, identifying properties, using complex patterns of speech)
- **Can children describe in words and drawings how they assembled their branch puzzles?**
  (comparing properties, identifying patterns and sequence, developing and using models)

Standards

Head Start Early Learning Outcomes Framework
P-SCI 1. Child observes and describes observable phenomena (objects, materials, organisms, events, etc.). *"Makes increasingly complex observations of objects, materials, organisms, and events. Provides greater detail in descriptions. Represents observable phenomena in more complex ways, such as pictures that include more detail."*
Next Generation Science Standards
Science and Engineering Practice: Developing and using models. *"Compare models to identify common features and differences. Develop and/or use a model to represent amounts, relationships, relative scales (bigger, smaller), and/or patterns in the natural and designed world(s)."*
Common Core State Standards for Mathematics
K.G.A.1. Identify and describe shapes. *"Describe objects in the environment using names of shapes, and describe the relative positions of these objects using terms such as above, below, beside, in front of, behind, and next to."*
Common Core State Standards for English Language Arts
SL.1.1. Comprehension and collaboration. *"Participate in collaborative conversations with diverse partners about grade 1 topics and texts with peers and adults in small and larger groups."*

Adopt a Tree

with Lauren M. Shea

Lesson: Observing a selected tree and comparing observations over time

Learning Objectives: In this ongoing unit of study, children will use their observation and comparison skills to understand that certain trees change throughout the seasons.

Materials: A tree, paper bags, a camera, and drawing and writing materials

Safety: This lesson will involve being outside. As with all field trips, scout out the area before you take children there; make sure the area is safe and free from hazards. Be careful to stay away from poisonous plants. Be aware of children in your class who may have allergies (pollen, bees, etc.) and plan accordingly. Also, be mindful of roots, obstructions, and other hazards along your path outside. It will be helpful to have additional adults to help supervise children while you are outside.

Teacher Content Background: Leaves contain several different compounds that absorb light and aid in photosynthesis. These compounds are different colors, including green, red, orange, and yellow. However, the most predominant of these compounds is chlorophyll. Because of the large amounts of chlorophyll in leaves, they appear green although other colors are present.

The leaves of *deciduous trees* change colors through the seasons. During the fall, these trees actively break down and reabsorb the chlorophyll for use the following spring. Without chlorophyll present, the colors of other pigments shine through, so leaves turn yellow, orange, or red before they drop from the trees. In the winter, the trees are bare. Examples of deciduous trees include maple, oak, elm, aspen, and cottonwood.

Evergreen trees retain their green coloring and foliage year-round. They do shed leaves as the leaves age, but the green color remains. There are two main types of evergreen trees: broadleaf and needleleaf. The broadleaf evergreens, such as eucalyptus and magnolia, are related to the deciduous trees and are similar in appearance. The needleleaf evergreens, such as pine, fir, cypress, and spruce, have leaves in the form of either bunched or single needles. (For more information about trees, see *Branch Puzzles*, p. 157.)

Science terms that may be helpful for teachers to know during this lesson include *observe, tree, leaf, trunk, branch,* and *seasons.*

Procedure

Getting Started

[Note: This lesson is an ongoing unit of study that will encourage children to investigate, explore, document, compare, and communicate findings over the course of the school year (including all four seasons: summer, fall, winter, and spring). Plan to initiate this lesson early in the school year and to revisit it regularly.]

Introduction: With your children, identify a "class tree." For the purposes of this lesson, it is recommended that you select a tree that

- is deciduous (with leaves that will change color, drop in the fall, and sprout anew in the spring);
- is clearly visible from top to bottom (not blocked by other objects);
- is nearby, and preferably visible from, your classroom; and
- has a safe area around it (children should be able to walk around the tree safely without danger from cars, low-hanging branches, obstructions, etc.).

If there are several trees near the classroom, ask children to choose one tree to be their special "adopted" tree. Children can vote and tabulate the results to determine the winner. Visit your adopted tree for the first time early in the school year so that children will be able to observe changes throughout the year.

Curiosity: Allow children to freely and safely explore your class tree and its surroundings. Encourage children to walk around the tree, look at the tree, touch the tree, listen to the leaves rustle, smell the leaves, and share anything they wonder about the "new addition" to their class. Name the tree and incorporate the tree into your class routine (read stories to children while sitting beneath your class tree and play games, such as Ring Around the Rosie, or Duck, Duck, Goose, with the tree at the center) to help children make an emotional connection to the tree and help to sustain children's interest throughout the year.

Investigating

Observing: Help develop children's oral language by modeling for them what you might think as you observe the tree. Say aloud to them, "When I observe this tree, I see it has green leaves. What do you see when you observe the tree?" and encourage children to use the sentence starter "When I observe the tree, I see ___" as they share their observations with a partner and then the group. Help children notice the many interesting attributes of the trunk, leaves, and branches. "What do you notice about the ground around the tree? What are the leaves like? What do leaves [bark, branches] feel like?" Some children may be interested in the leaves on the ground, whereas others may notice the sap or insects on the trunk of the tree. Follow children's leads in deciding which questions to ask.

To encourage additional observations, provide each child with a paper bag to bring along as you visit your class tree. Have each child collect from the ground two or three important items that best describe the tree in its current state. Choosing a limited number will encourage the children to make purposeful decisions about what they think are the most important parts of the tree. Use the children's collection bags to extend the conversation about your tree. Ask children, "What did you collect? What did you notice while visiting the tree?" Record some of the ideas on chart paper and ask children what they found in common. With your children, collaboratively review the observations and collected materials and have

Measuring a tree

children decide on important observations to record using photographs. Help each child use a camera to "collect" an observation of the tree. Or, if you'd prefer, children can make a drawing of the tree, or part of the tree, instead. Either way, be sure to date these observations. [Note: If the ground around your tree is free of fallen leaves, twigs, fruits, flowers, etc., skip the collection bags and just have children "collect" observations through drawings or photographs.]

Comparing and Measuring: A month later, return to the tree. Have the children observe the tree again as described in the *Investigating* section. Encourage children to talk about their initial observations of the tree. Next, discuss the differences in the tree. Ask children to compare the tree now to their earlier drawings or photographs. Use questioning to help children notice if anything has changed about the tree. "What did you notice? How has the tree changed? How have the tree's surroundings changed? How do the leaves look now compared to the last time we visited?" Encourage children to share their observations of the tree. If you took measurements of the tree, repeat the same measurements again (see *Math Connections*). Help the children compare the measurements for any changes in the size of the tree or its leaves. Finally, have children complete another drawing or photograph of the tree, again making sure to include the date on the drawing. Continue to visit the tree monthly. With each visit, help the children draw or photograph the tree and compare the tree to earlier observations, describing any changes they notice.

Making Sense

Application: Just as scientists apply their current knowledge to begin to question, investigate, and understand new situations, children in your class might start observing other plants and trees and begin to wonder about other plants' leaves changing colors, or not changing, across the seasons. If needed, prompt your children to compare their adopted tree to other plants, including evergreen plants. Listen to your children's ideas and encourage them to make additional observations and to offer explanations for the differences they see. [Note: Supporting children as they apply ideas learned in one context to a new, related context is important for developing critical-thinking skills and meaningful, long-term learning.]

Generating Explanations: Throughout the school year, children will have observed several changes to the tree, including changes to its environment and leaves. Near the end of the school year, review the drawings or photographs recorded throughout the year. As the children review their class tree throughout the seasons, they may begin to notice that there is an annual cycle that the tree goes through. With recorded observations (drawings or photographs) in front of them, challenge children to sequence the observations and then ask children to predict what the tree will look like after summer vacation is over, next winter, this time next year, and so on. Encourage children to refer to observations as they predict and again as they explain their thinking. As they predict, don't emphasize the "correct" answers; keep in mind that the goal is to provide children opportunities to explain their thinking and to develop their reasoning skills.

What's Next?

Extension Activity

Initiate a conversation about the benefits of trees by asking children how trees help people (possible responses may include tree swings, shade, fruits or nuts to eat, climbing, firewood, etc.). Ask children to brainstorm different foods that come from trees. Share and sample with children different foods that come from trees and graph which ones are their favorites. Share examples of different foods and have children sort foods into two groups: "food from trees" and "food not from trees." Provide children with old magazines and let them cut out pictures of foods that grow on trees and foods that do not grow on trees. Children can glue these onto a chart.

Ask children if they can think of things that are made from trees. Wood and paper are made from trees. Prompt children to brainstorm different things made of wood and things made of paper. Invite your children to go on a "things made from trees" scavenger hunt—searching for, and sharing with their classmates, tree products found in your classroom. Share a book such as *When Dad Cuts Down the Chestnut Tree* by Pam Ayres and Graham Percy (1988), *A Tree Is Nice* by Janice Urdy (1987), *Be a Friend to Trees* by Patricia Lauber (1994), or *Thank You, Trees! (Tu B'Shevat)* by Gail Langer Karwoski (2013) to expand children's thinking about the different ways that trees benefit people.

Integration to Other Content Areas

Reading Connections

In each part of this ongoing lesson, there are many opportunities to include reading and literature. Each time you have the children draw, document findings, and record their words, you are promoting the idea that printed words carry meaning. After each visit to the tree, fill your classroom library with books about trees. Let children have free time to explore the books, look at the pictures, and talk about what they see. Conduct readalouds as an opportunity for children to connect literature to their real-life experiences at the tree. When children read or hear something about a tree that they have discovered in their own investigations, they will feel like experts and it will motivate them to learn even more. Several books, including *Fall Leaves: Colorful and Crunchy* by Martha Rustad (2011), *Tree: A Peek-Through Picture Book* by Britta Teckentrup (2016), *Sky Tree* by Thomas Locker (2001), *The Seasons of Arnold's Apple Tree* by Gail Gibbons (1988), *Fall Leaves Fall* by Zoe Hall (2000), *Tree for All Seasons* by Robin Bernard (2001), and *Tap the Magic Tree* by Christie Matheson (2013), have pictures and text that will help children explore the seasons and the seasonal changes of some trees. (See *Branch Puzzles*, p. 157, for additional books about trees.)

Writing Connections

Encouraging children to record details as they document their observations of the tree over time promotes their writing development. Support their letter formation and sound recognition as they document and record. If children are ready for initial sound–letter correspondence, you may encourage them to label the parts of the tree. Engage in a group writing activity after each tree observation—have the children help you create the bulletin board with their work. Ask for children's input about the design and content of the bulletin board; include what they think is important to show about your class tree. Write each child's name next to an observation he or she made while visiting the tree to show the children that their oral language can be translated into written print.

Math Connections

While visiting the tree, children can measure the circumference of the trunk with a piece of string or, using a longer string, measure the length of the tree's shadow. Then they can compare the size of the class tree to other trees in the area. Children can also compare the length of the string to their own bodies by placing the string around themselves or along their shadows. For the circumference, align the string along a yardstick to determine the circumference in standard units (inches or centimeters). Although the trunk is unlikely to change noticeably during the school year, the leaves will. Have children use standard units to measure the length and width of several leaves and to compare these with leaf measurements taken earlier in the year. Another way to incorporate math skills is with tallying. Using the children's collection bags (or photographs), tally the number of each type of item that was collected. Graph the number of items in each group. Also, children can sort the collected items in many different ways (by color, size, texture, etc.) and then chart the number of items in each group.

Other Connections

Child's Life Connection

The seasonal nature of many plants makes them ideal for learning about patterns, change over time, and sequence. Give children pictures of your class tree taken during different seasons. Children can work together to sequence the pictures of the tree and explain their thinking. [Note: Keeping these pictures year after year will give your future children opportunities to see the growth of your class tree over time.] Of course, there are many other things in children's lives that grow and change over time. Ask families for four or five pictures of your children, one picture each of different stages in their child's life (newborn, infant, early toddler, late toddler, and recent). Children will enjoy viewing, describing, and sequencing pictures of themselves and their classmates.

Center Connections

There are many ways to explore and think about trees in the classroom. In the **art center**, an adult can guide children in making a paper bag tree. Have children place a piece of cardboard in the bottom of the bag and then, a few inches from the bottom, twist the bag several times. Cut four or five slits from the top of the bag down to the twist. Twist each of the resulting strips to form branches. Children can glue pieces of tissue paper to the branches to represent leaves. Encourage children to select tissue paper colored the same as the leaves on your tree or to make trees that represent the different seasons. Objects collected from the tree can be placed at the **sensory table**. Children can feel, touch, manipulate, and sort the real objects from their tree. Providing materials from two different trees can lead children to compare the two, noticing differences in how the parts of different trees look, feel, and even smell. Your **dramatic play** area can be turned into a tree farm or nursery where the children can take care of trees. Provide artificial trees and everything children would need to care for, buy, and sell the trees. Consider watering cans, gloves, overalls, stuffed animals that live in trees, cash registers, and more.

Family Activities

Your child has been learning about trees and how some trees change across the four seasons. You can help your child learn more about trees. Take your child outside to count and observe the trees that are growing in your yard or neighborhood. Select one tree and, once a month, observe this tree with your child. Ask your child to identify and describe anything she or he notices about the tree. Share your own observations with your child. Talk together about the size, shape, color, scent, and textures of the tree. Look for fruit, leaves, seeds, bark, or twigs that may have fallen from the tree. Spend time comparing these fallen parts of the tree with the same parts that are still on the tree. Ask your child questions about the tree, including "What do you notice about the tree? What color are the leaves? How big is the tree? How is it similar to or different from the trees around it?" Share your interest in trees and have fun exploring trees together.

Actividades Familiares

Su hijo ha estado aprendiendo sobre los árboles y cómo algunos cambian a través de las cuatro estaciones. Puede ayudar a su hijo a aprender más sobre los árboles. Lleve a su hijo de paseo a contar y observar los árboles que crecen en su jardín o en el vecindario. Elija un árbol y, una vez al mes, obsérvelo con su hijo. Pídale a su hijo que identifique y describa cualquier cosa que note alrededor delsobre el árbol. Comparta sus propias observaciones con su hijo. Hablen juntos sobre el tamaño, forma, color, aroma y textura del árbol. Busquen frutas, hojas, semillas, corteza o ramas que pueden haber caído del árbol. Pase tiempo comparando estas partes caídas del árbol con las mismas partes que aún se encuentran en el élárbol. Haga preguntas sobre árbol, incluyendo, "¿Qué puedes observar sobre el árbol? ¿De qué color son las hojas? ¿Qué tan grande es el árbol? ¿Cómo se asemeja similar o diferentecia de los árboles a tu alrededor?" Compartan su interés en los árboles y diviértanse explorando los árboles juntos.

Assessment–What to Look For

- **Can children describe the class tree based on their observations?**
  (using sense to gather information, identifying properties, using complex patterns of speech)
- **Can children record (draw or write) observations and contribute to class discussion?**
  (using new or complex vocabulary, documenting and reporting findings, discussing scientific concepts, listening to and understanding speech)
- **Can children observe and describe changes in the tree throughout the seasons?**
  (comparing properties, identifying patterns and change, using complex patterns of speech, using new or complex vocabulary)

Standards

Head Start Early Learning Outcomes Framework
P-SCI 6. Child analyzes results, draws conclusions, and communicates results. "Analyzes and interprets data and summarizes results of investigation. Draws conclusions, constructs explanations, and verbalizes cause and effect relationships."
Next Generation Science Standards
Science and Engineering Practice: Constructing explanations and designing solutions. "Make observations (firsthand or from media) to construct an evidence-based account for natural phenomena."
Common Core State Standards for Mathematics
K.MD.A.2. Describe and compare measurable attributes. "Directly compare two objects with a measurable attribute in common, to see which object has 'more of'/'less of' the attribute, and describe the difference."
Common Core State Standards for English Lanuage Arts
W.K.3. Text types and purposes. "Use a combination of drawing, dictating, and writing to narrate a single event or several loosely linked events, tell about the events in the order in which they occurred, and provide a reaction to what happened."

Comparing Leaves

with Nicole Hawke

Lesson: Collecting, observing, and comparing the different sizes, shapes, and colors of leaves

Learning Objectives: Children will use their observation skills to closely examine leaves. Children will compare leaves, noticing that leaves can vary in size, shape, and color.

Materials: Leaves from many different kinds of trees, drawing and writing materials, bags for collecting leaves

Safety: This lesson will involve being outside. As with all field trips, scout out the area before you take children there; make sure the area is safe and free from hazards. In particular, look carefully for poison ivy, poison oak, stinging nettle, and other plants that can cause severe skin irritations. Be aware of children in your class who may have allergies (pollen, bees, etc.) and plan accordingly. Also, be mindful of roots, obstructions, and other hazards along your path while you are all outside. It will be helpful to have additional adults to help supervise children while outside. At the end of this lesson, be sure children thoroughly wash their hands with soap and water.

Teacher Content Background: Leaves are usually green in color because their cells are filled with billions of chlorophyll molecules. These relatively large—and vitally important—molecules reflect green light, but absorb other wavelengths of light and, in doing so, harness energy from light and convert that energy to a form useful for metabolism and growth (i.e., chemical energy). Within the leaves, the energy captured by chlorophyll is used to recombine the atoms in water (H_2O) and carbon dioxide (CO_2) to form oxygen (O_2) and a unit (CH_2O) used in a molecule of glucose ($C_6H_{12}O_6$), a simple sugar used to power the different activities of plant cells. Because the winter has fewer daylight hours and colder temperatures that often freeze water, making it unavailable to plants, during winter photosynthesis decreases and leaves become inefficient. In anticipation of this, many plants have evolved to eliminate their leaves in the fall and to live off the glucose they have stored for the winter. But prior to dropping their leaves, plants break down and reabsorb the chlorophyll, revealing yellow, orange, and red colors that are created by molecules other than chlorophyll that absorb different wavelengths of light. These colors were present in the leaf all along but were drowned out by the much more prevalent green chlorophyll. (See *Terrariums*, p. 101, and *Adopt a Tree*, p. 164, for additional information.)

Science terms that may be helpful for teachers to know during this lesson include *investigate*, *observe*, *tree*, *leaf*, *photosynthesis*, and *chlorophyll*.

Procedure

[Note: This lesson encourages children to look at many different characteristics of leaves, including color. You may choose to teach this lesson in the fall when there is a greater variety of leaf colors available. As you begin this lesson, be sure you and your children have access to a variety of trees.]

Getting Started

Prior Knowledge: Share with your children that they will be studying plants, and use questioning to probe children's prior knowledge about plants and leaves. Have children think about the plants they have seen, and ask them to share the different parts of a plant and the different colors they have seen on plants. Children will likely identify and describe leaves, among other plant features.

Initial Explanation: Once children have named several plant parts, focus their attention on leaves. Ask them about the different colors, shapes, and sizes of leaves they have seen and make a list of the descriptions children provide. As you record the children's describing words, provide picture support or other cues next to each word (e.g., if they say leaves are green, use a green marker to write the word *green*). After generating a class list, have each child draw his or her own picture of a leaf. Do not provide prompts about how the leaves should look in color, size, or shape. Ask children to show and describe their drawing to a partner.

Investigating

Observing: Now that children have recorded their initial ideas, take them for a "leaf walk" around your schoolyard to make observations of leaves. Give each child a small bag to collect leaves and encourage children to select leaves that are very different from each other. Limiting children's collections to just three leaves will encourage them to be thoughtful and selective as they collect. [Note: If your school grounds don't have an abundance and variety of plant life, it's a good idea to bring various leaves to distribute to children during your leaf walk. Also, additional adults may be needed to help children reach leaves.]

Comparing and Sorting: Back in the classroom, have children spread their leaves out in front of them. Ask children to look closely at the leaves and then to tell you words that describe each leaf. While children are examining and describing the leaves, ask the children guiding questions, such as "In what ways are these leaves different? How are they similar? Which plant did they come from? What do you know about leaves?" Then have the children compare, sort, and classify their leaves in numerous different ways, guiding children to specifically look at the different sizes, colors, and shapes of leaves.

Making Sense

Describing Findings: Have children revisit and compare their initial drawings to actual leaves, looking for leaves that are similar to their drawings and leaves that are different from their drawings. Help promote children's oral language development and critical-thinking skills by asking them to explain their choices. Use questions such as "Which leaf is most like your drawing? What are some similarities that this leaf and your drawing have? Show me a leaf that is very different from your drawing. What are some differences this leaf has compared to your drawing?" Encourage children to talk with each other; have them work in pairs or small groups to share their leaves. During this sharing, children can explain how they sorted their leaves by color, shape, and size; describe the leaves that are most similar to and different from their drawings; and discuss which leaves came from the same trees.

Curiosity: Ask children if, after observing their leaves, they have any new questions about leaves. If children were able to find a tree that had both green and different-colored leaves, ask the children why they think the tree has different-colored leaves and if they think the leaves change color or not and why. Show any leaves that you found that were nibbled on by animals and ask the children why they think these leaves look different. Throughout this conversation, listen to and validate children's explanations and model enthusiasm and wonder for nature—helping to increase children's interest in and amazement at the things they observe in nature.

What's Next?

Extension Activity

The perfect follow-up activity is to help children create their own leaf collections. Children can create little books with different leaves glued to each page. Consider the different attributes of leaves; children can collect and create a book of leaves of different sizes (big, small, narrow, wide, etc.); colors (green, yellow, red, etc.); or other characteristics (fuzzy, jagged, pointy, bumpy, etc.). Or if you have many trees and other plants at your school, create a class book of "The Leaves at My School." Each child can choose a plant to photograph and collect a leaf from. Children can each create a page of the class book with a photograph, a leaf, and a description of a plant. For leaf collections, there are many ways to preserve leaves. The easiest is to simply press and dry the leaves. Place the leaf between two paper towels on a hard surface and place a heavy book or two on top of the leaf. In two or three days the leaf will be ready to be included in your collections. [Note: Preserving leaves helps to maintain their flattened shape and, depending on the technique you use, their color as well. If you glue an unpreserved leaf to a page, as it dries it will curl, lose its shape, and pull away from the page.]

Integration to Other Content Areas

Reading Connections

Early literacy skills can be supported at various points throughout this lesson. Books can be used to further develop these skills while expanding children's interest in and understanding of leaves. There are several informational books about different kinds of leaves, including *Why Do Leaves Change Color?* by Betsy Maestro (2015), *Autumn Leaves* by Ken Robbins (1998), *My Leaf Book* by Monica Wellington (2015), and *Fall Walk* by Virginia Brimhall Snow (2013). Reading these books and having them available in your classroom library will encourage children to use books as additional sources of knowledge and

Tracing a leaf

information. Also, sharing books like *Leaves* by David Ezra Stein (2010), *Leaf Trouble* by Jonathon Emmett (2009) or *The Little Yellow Leaf* by Carin Berger (2008) can show children that stories related to their science observations can be read for enjoyment.

Writing Connections

Leaf collections, as described in the extension activity, provide great opportunities for children to write about leaves. For each leaf collected, children can write (with your assistance) a description of the leaf and of the plant the leaf came from. Children can also write stories about the leaf walk or about interesting leaves they found.As you and the children write and read these sentences together, ask children to demonstrate different early literacy skills such as, "Find a [*w*] that sounds like [/w/]. Point to the word *tree*. Put your finger on a space, under a period, at the beginning of a sentence."

Math Connections

Children can use several different math practices as they explore their leaves. While sorting the leaves, children can practice with words related to size as they arrange leaves from largest to smallest, longest to shortest, widest to narrowest, and so on. Additionally, children can quantify these sizes using standard and nonstandard units as they measure their leaves. For another connection to math, use leaves to search for shapes. Many leaves come in diamonds, triangles, ovals, and other shapes. With younger children, cut the leaves into shapes for children to examine. Have children identify and describe the characteristics of each shape.

Other Connections

Child's Life Connections

Help children to see the possibilities for art in their everyday lives. For inspiration, share books with children like *Leaf Man* by Lois Ehlert (2005) and *Look What I Did With a Leaf!* by Morteza Sohi (1993) that show different objects made from leaves. Have children bring leaves from their homes or neighborhoods to school and allow children to arrange their leaves to form different shapes. Encourage children to describe their art to you. Children can glue the leaves to a piece of paper to keep them in place and with your help write the title of their creation.

Center Connections

There are many ways to bring learning about leaves into the classroom. In the **art center,** children can make leaf rubbings. Help children tape their leaf down on a hard surface, with the underside of the leaf facing up, and then place a white sheet of paper on top. Children can select a crayon that is the same color as their leaf and then, using the side of the crayon, can lightly color over the leaf on the white paper. Assemble children's rubbings in a whole-class leaf collage. At the **sensory table,** put out a large collection of leaves. Provide children with containers that they can use to sort and place the leaves in. Also provide tongs, tweezers, and other grasping and scooping tools that will help the children build their fine motor skills as they find different ways to pick up and hold the leaves. For **dramatic play,** consider setting out some leaves (larger leaves work best). Children can hold the leaves and pretend they are trees, acting out how trees sway in the wind and the sounds made as wind rustles through the tree's leaves. Children can also act out the life of a tree, growing taller and wider and sprouting leaves, dropping leaves, and sprouting them again while other children pretend to care for the "trees."

Family Activities

At school, your child has been exploring leaves. Foster your child's interest in science by engaging in several different leaf activities. Go on a leaf walk with your child. See what kinds of leaves you both can find. During your walk, encourage your child to talk about the leaves he or she finds and to describe the color, shape, and size of leaves. Make leaf art together. Near the bottom of a large piece of paper, have your child draw the trunk of a tree. Above this you and your child can glue the leaves you've collected during your leaf walk to make a tree. Eat leaves together. Incorporate leafy vegetables into your meals. Try spinach, chard, bok choi, turnip, collard greens, mustard greens, or a salad with different kinds of romaine and iceberg lettuce. Discuss the types of leaves you eat and go to the library or on the internet to find pictures of the plants growing. Be sure to remind your child that only the leaves that come from the grocery store are safe to eat.

Actividades Familiares

En la escuela, su hijo ha estado investigando acerca de las hojas. Fomente el interés de su hijo en la ciencia participando en distintas actividades en relación con las hojas. Lleve a su hijo a dar un paseo en busca de hojas. Vea qué tipos de hojas ambos pueden encontrar. Durante el recorrido, anime a su hijo a hablar sobre las hojas que encuentrea y a describir el color, la forma y el tamaño de las hojas. Hagan obras de arte con las de hojas juntos. Cerca de la parte inferior de una hoja grande de papel, pídale a su hijo que dibuje el tronco de un árbol. Por encima de esta, usted y su hijo pueden pegar las hojas que ha recogido durante su paseo para hacer un árbol. Coman hojas juntos. Incorpore las verduras de hojas de verduras en sus comidas. Pruebe las espinacas, acelga, repollo chino, hojas de nabo, acelga, y/o de mostaza; o una ensalada con distintos tipos de lechuga romana y lechuga de bola. Discuta los tipos de hojas que consume y vaya a la biblioteca o al Internet para encontrar fotos de las plantas en crecimiento. Asegúrese de recordarle a su hijo que sólo las hojas que se comprar en el supermercado o almacén local son seguras para comer.

Assessment—What to Look For

- **Can children describe leaves based on their observations?**
  (using senses to gather information, identifying properties, using complex patterns of speech)
- **Can children sort leaves based on observable characteristics?**
  (identifying properties, comparing properties, classifying)
- **Can children explain their reasoning?**
  (constructing explanations, explaining based on evidence, using complex patterns of speech)

Standards

Head Start Early Learning Outcomes Framework
P-SCI 1. Child observes and describes observable phenomena (objects, materials, organisms, events, etc.). "Makes increasingly complex observations of objects, materials, organisms, and events. Provides greater detail in descriptions. Represents observable phenomena in more complex ways, such as pictures that include more detail."
Next Generation Science Standards
Science and Engineering Practice: Analyzing and interpreting data. "Record information (observations, thoughts, and ideas). Use and share pictures, drawings, and/or writings of observations. Compare predictions (based on prior experiences) to what occurred (observable events)."
***Common Core State Standards* for Mathematics**
K.G.A.1. Identify and describe shapes. "Describe objects in the environment using names of shapes, and describe the relative positions of these objects using terms such as *above, below, beside, in front of, behind,* and *next to.*"
***Common Core State Standards* for English Language Arts**
L.K.5.A. Vocabulary acquisition and use. "Sort common objects into categories (e.g., shapes, foods) to gain a sense of the concepts the categories represent."

Scavenger Hunt

with Lauren M. Shea

Lesson: Observing, comparing, and classifying objects in nature

Learning Goals: Children will use their observation skills (particularly using their senses of sight and touch) to find objects with similar characteristics while on an outdoor nature scavenger hunt. Children will compare the texture, size, shape, and color of different objects from nature.

Materials: Two large leaves with different colors and textures; a collection of twigs, rocks, leaves, flowers, seedpods, and other materials easily found in your outdoor environment; magnifiers; paper bags

Safety: This lesson will involve being outside. As with all field trips, scout out the area before you take children there; make sure the area is safe and free from hazards. In particular, look carefully for poison ivy, poison oak, stinging nettle, and other plants that can cause severe skin irritations. Be aware of children in your class who may have allergies (pollen, bees, etc.) and plan accordingly. Also, be mindful of roots, obstructions, and other hazards along your path outside. Additional adults may be needed to aid children in reaching leaves and helping to ensure their safety. At the end of this lesson, be sure children thoroughly wash their hands with soap and water.

Teacher Content Background: In popular use, the word *observe* means to look at. However, in science, making observations is much more than that. Observing is the process by which scientists collect data. Science observations can involve sight but are not limited to that single sense. Instead, scientists use all of their senses and frequently use tools to help extend and increase the sensitivity of their senses. Thermometers, magnifiers, balances, telescopes, x-ray machines, oscilloscopes, and spectrometers are all tools that help scientists to "sense" the natural world with great precision. Interpretations of observations help scientists to generate explanations. Typically, scientists state their explanations, based on observations and other science information, in a manner that is testable. These scientifically sound, testable explanations of observed phenomena are called hypotheses. Experimental testing of a hypothesis provides new observations that, when analyzed, can support, refute, or generate altogether new explanations. This ongoing process of observing and testing is a cornerstone of science.

Science terms that may be helpful for teachers to know during this lesson include *observe, compare*, and *characteristic*.

Procedure

[Note: For this lesson, you'll need to collect several natural objects that can be found in your schoolyard (or local park), such as leaves, sticks, rocks, flowers, etc. Collect multiple examples of each object (three of the same type of leaf, flower, etc.) so that you have three similar sets. One set of objects will be shared and posted during the *Getting Started* section, another set will be distributed during the *Investigating* section, and the third set will be used during the *Making Sense* section.]

Getting Started

Prior Knowledge: To begin this investigation, ask the children what they should do when they observe and compare objects. If needed, ask more focused questions such as "What are some of the characteristics to look for when observing and comparing? Which senses do you use to make observations? How could your sense of touch (sight, smell, etc.) help you to observe?" The children might talk about using their eyes and hands to observe shape, size, texture, and color. Show the children two large leaves that differ in color, texture, and other characteristics and ask them to use their senses of sight and touch to observe and compare the leaves. Have children share one difference with a partner sitting next to them and then present their thinking to the class.

Introduction: Share with children the different nature objects that you have collected. Give children magnifiers and ask them to look closely, making detailed observations as they describe the characteristics (color, texture, shape, flexibility, smell, etc.) they notice about the objects in your collection. Tape a sample of each object to a poster board and, next to each sample, write the name of each object, highlighting the first letter for initial sound correspondence (e.g. ***L****eaf*, ***S****tick*, ***R****ock*). Beneath each name, using words and picture support, record the descriptions children offer for each object. Take the poster outside with you to serve as a reference for the children during the next part of the lesson.

Investigating

Observing and Comparing: Have children select one object from your collection. Tell children that they will go outside and attempt to find other objects in nature that are similar to their object and let them know that they'll have to find objects that are the same size, color, shape, and texture as the object they've selected. Choose an object yourself and express your thoughts out loud as you model this process for the children (e.g., "This rock is brown. I wonder what other things I can find outside that are also brown. Maybe a leaf. Sometimes leaves can be brown. And this rock is smooth; what other things can I find that are smooth like my rock?"). When your class is ready, give children small bags to collect objects in and take the children to the schoolyard or to a local park. Have children search for one characteristic at a time. For example, first instruct children to find something that has the same texture as their object and, after they've collected something with a similar texture, ask them to find something that is similar in color. Encourage and guide children as they search, asking questions such as "Describe the texture of your rock; how does it feel? Can you find something that has the same texture as your rock?" Your children will enthusiastically share their findings with you. Demonstrate your own enthusiasm for science and ask children to share more about their discoveries. "Wow! What did you find?! That's great; which of its characteristics are similar to your object?" As children search, encourage them to help each other, discuss their findings together, and compare the objects they find with the ones on the poster from the *Getting Started* section of the lesson.

Collecting objects from nature

Making Sense

Describing Findings: Once children have found similar objects for each characteristic, return to the classroom to discuss your findings. First, let children share and describe their discoveries in small groups or with a partner. As they share, encourage them to compare their different objects (e.g., "How is the rock you collected different from the one your partner collected?"). Then, for each characteristic, choose an object from your collection and have children search their bags to find an object that is similar. Use questions to encourage children to describe their thinking (e.g., "Size describes how big something is. Do you have an object that is similar in size to this leaf? Describe how your object is similar to my leaf." "Texture describes how something feels. Do you have an object that is similar in texture to this rock? Describe how your object is similar to my rock."). Discuss with children how different objects can have similar characteristics (sticks and rocks can be the same color) and how similar objects can have different characteristics (some leaves are long and some are short). Use the collected objects to show examples of this to the children, and encourage children to show you examples as well.

What's Next?

Extension Activity

After the lesson, you can extend your children's learning about the different characteristics of natural objects by engaging children in a sorting activity. Set out several bins, one for each specific characteristic studied during the lesson and labeled, using words, picture support, and an example or two from your collection. For example, you could label a bin with the word *bumpy*, a drawing of a bumpy surface, and a bumpy object such as a rock or a piece of bark. Children can sort their items into the different bins and discuss their thinking as they decide where to place objects that have characteristics of multiple bins. Deliberations over whether to place their pine needles into the "green," "pointy," or "long" bin can lead to great thinking and language use. This sorting can inspire new scavenger hunts: "There are no objects in our 'red' bin. Let's go on a scavenger hunt for red things" and can be repeated by children interested in sorting objects by different characteristics.

Integration to Other Content Areas

Reading Connections

There are many books that relate to scavenger hunts. When children are unable to get outdoors, they can interact with "search and find" books to mimic their scavenger hunting. Choose books with natural themes such as *1001 Bugs to Spot* by Emma Helbrough (2009), *1001 Animals to Spot* by Rith Brocklehurst (2010), or *Let's Find It: My First Nature Guide* by Katya Arnold (2002). There are many fun stories about searching for objects from nature, including several that are all variations of the familiar children's chant, "Can't go around it. Have to go through it" including *We're Going on a Leaf Hunt* by Steve Metzger (2008), *We're Going on a Lion Hunt* by David Axtell (2015), and *We're Going on a Bear Hunt* by Michael Rosen (1997). And there are many books that include popular characters going on scavenger hunts, including Dora the Explorer, Peppa Pig, Winnie the Pooh, and many others. You may also want to share books on the five senses and help children to consider all the different types of characteristics they can observe. (See *Smelling Plants*, p. 109, for a list of books about the senses.)

Writing Connections

After the science lesson, engage children in a shared writing activity to summarize their findings. For the shared writing, you can write a sentence that describes your scavenger hunt or one that lists the nature objects found during your hunt as children come up and help you to write certain letters, hold a space marker, or draw a punctuation mark. This summary sentence may be read and reread at morning meeting so children can revisit their comparisons of objects found during their scavenger hunt. Scavenger hunts can also provide opportunities to promote letter recognition. Share books such as *Discovering Nature's Alphabet* by Krystina Castella and Brian Boyl (2006)and *Alphabet City* by Stephen Johnson (1999) to inspire you and your children to search for and photograph letters in nature and around your schoolyard. Back in the classroom, print and review the photographs to identify the letters again.

Math Connections

This lesson provides opportunities to develop several important skills as children match, sort, put in series, and regroup the objects according to one or two attributes such as color, shape, or size—give children continued opportunities to practice these skills. Scavenger hunts also provide additional opportunities for math. Share books such as *Star in My Orange: Looking for Nature's Shapes* by Dana Meachen Rau (2006), *City Shapes* by Diana Murray (2016), *Find Spot!* by Stacey Previn (2014), *Find the Triangle* by Britta Teckentrup (2016), and *What Shape Is It?* by Bobbie Kalman (2007) to inspire you and your children to take "shape scavenger" hunts. In addition to shapes, children can hunt for amounts (e.g., searching the playground to find one sandbox, two slides, three swings, etc.). As children go on these math-based scavenger hunts, have them take photographs of their findings. Children will enjoy hunting for shapes and amounts and sharing their discoveries.

Other Connections

Child's Life Connection

As children become more interested in searching nature, encourage them to try other nature scavenger hunts. Ask them to think about any other objects from nature that can be found, collected, observed, and compared. Generate with children a list of different attributes and types of things that can be searched for. After completing this lesson, children should be able to think of many possibilities, such as hunting for different shapes, flowers, or animals or hunting for things that are yellow, fuzzy, or soft. Give children the autonomy to decide what their next search should be and encourage them to conduct their next "scavenger hunt" on their own or with a friend.

Center Connections

The idea of scavenger hunting is very exciting to young children. In your **art center**, children can use the objects they found during the lesson to create a unique art project. They may choose to glue, paint, color, or sculpt with their findings. After everyone finishes their project, children can compare projects and discuss how they all used objects with similar characteristics to create different art. At the **sensory table**, the children can look at and touch some found objects to sort, match, or compare based on characteristics. Provide magnifiers and guide children by asking questions such as "What do you notice?" and "How are these objects similar?" In the **dramatic play** area, the children can be nature trackers or scientists searching for their particular objects of study. The teacher may hide certain objects in the dramatic play area so children can pretend to be geologists (rocks/minerals), botanists (leaves/plants), or paleontologists (fossils/bones). The children can compare the different characteristics of the objects they find.

Family Activities

At school, your child has gone on scavenger hunts, searching for objects from nature with common characteristics. She or he searched for objects that were similar in color, texture, size, and shape. You can continue this scavenger hunt at home! This time, your child can pretend to be the teacher and you can find a matching object in and around your home. Your child may instruct you to find something that is the same shape as a ball or something that is the same color as a leaf. Have your child compare her or his object with the one you find (for example, comparing her or his ball and the apple you found or a leaf and the green crayon you found). As you compare together, consider: Is your object bigger? Smaller? A different color? The same size? Is its texture different? How does it feel? Switch roles and give your child a different characteristic to search for. Talk about each comparison and play again and again!

Actividades Familiares

En la escuela, su hijo ha salido en búsquedas del tesoro para buscar objetos de la naturaleza con características comunes. Buscaron objetos que eran similares en color, textura, tamaño y forma. ¡Puede continuar con esta búsqueda del tesoro en casa! Esta vez, su hijo puede jugar a ser el maestro y usted puede encontrar un objeto similar en y cerca alrededor de su casa. Su hijo puede indicarle que encuentre algo que tenga la misma forma que una pelota o algo que sea del mismo color que una hoja. Que suPregúntele a su hijo que compare su objeto con el que usted encontró (por ejemplo, comparando su bola pelota y la manzana que usted ha encontrado o su hoja y el lápiz de color verde que usted encontró). A medida que comparan juntos, tenga en cuenta: ¿Su objeto es más grande? ¿Más pequeño? ¿Tiene un color distinto? ¿El mismo tamaño? ¿Su textura es diferente? ¿Cómo se siente? Cambie de roles e indíquele a su hijo una característica diferente para que busque. ¡Hable de cada comparación y juegue una y otra vez!

Assessment—What to Look For

- **Can children describe objects from nature based on their observations?**
  (using senses to gather information, identifying properties, using complex patterns of speech)
- **Can children compare and sort objects based on observable characteristics?**
  (identifying properties, comparing properties, classifying)
- **Can children explain their reasoning?**
  (constructing explanations, explaining based on evidence, using complex patterns of speech)

Standards

Head Start Early Learning Outcomes Framework
P-SCI 3. Child compares and categorizes observable phenomena. "Categorizes by sorting observable phenomena into groups based on attributes such as appearance, weight, function, ability, texture, odor, and sound."
Next Generation Science Standards
Science and Engineering Practice: Planning and carrying out investigations. "Make observations (firsthand or from media) and/or measurements to collect data that can be used to make comparisons. Use observations (firsthand or from media) to describe patterns and/or relationships in the natural and designed world(s) in order to answer scientific questions and solve problems."
Common Core State Standards for Mathematics
K.G.A.1. Identify and describe shapes. "Describe objects in the environment using names of shapes, and describe the relative positions of these objects using terms such as above, below, beside, in front of, behind, and next to."
Common Core State Standards for English Language Arts
L.K.5.A. Vocabulary acquisition and use. "Sort common objects into categories (e.g., shapes, foods) to gain a sense of the concepts the categories represent."

Appendix A

Lesson Connections to the Head Start Early Learning Outcomes Framework

Lesson	**P-LIT-2.** Child demonstrates an understanding of how print is used (functions of print) and the rules that govern how print works (conventions of print).	**P-LIT 3.** Child identifies letters of the alphabet and produces correct sounds associated with letters.	**P-LIT 6.** Child writes for a variety of purposes using increasingly sophisticated marks.	**P-MATH 3.** Child understands the relationship between numbers and quantities.	**P-MATH 4.** Child compares numbers.	**P-MATH 5.** Child associates a quantity with written numerals up to 5 and begins to write numbers.	**P-MATH 8.** Child measures objects by their various attributes using standard and non-standard measurement. Uses differences in attributes to make comparisons.	**P-MATH 9.** Child identifies, describes, compares, and composes shapes.	**P-SCI 1.** Child observes and describes observable phenomena (objects, materials, organisms, and events).	**P-SCI 2.** Child engages in scientific talk.	**P-SCI 3.** Child compares and categorizes observable phenomena.	**P-SCI 4.** Child asks a question, gathers information, and makes predictions.	**P-SCI 5.** Child plans and conducts investigations and experiments.	**P-SCI 6.** Child analyzes results, draws conclusions, and communicates results.
Animals														
Roly-Polies	•	•	•	•	•	•	•		•	•	•	•	•	•
Jumping Crickets	•		•		•		•		•	•	•	•		•
Observing Earthworms	•		•	•	•		•		•	•	•	•	•	•
Snails	•		•		•		•	•	•	•	•	•	•	•
Swimming Fish	•		•	•					•	•		•		•
Critter Camouflage			•	•					•	•		•		•
Spiderwebs			•	•		•		•	•	•	•			
Feeding Birds			•	•	•	•			•	•	•	•	•	•
Plants														
Sorting Seeds			•	•	•			•	•	•	•			
Seeds in Our Food	•	•	•	•	•	•			•	•	•	•		•
Where Vegetables Come From	•	•	•				•		•	•	•	•		•
Sprouting Seeds	•		•	•	•		•		•	•	•	•	•	•
Terrariums	•		•	•	•		•		•	•	•			
Smelling Plants			•		•				•	•	•			
Pumpkin Outsides			•	•	•	•	•		•	•	•			
Pumpkin Insides		•	•	•	•				•	•	•			

Lesson Connections to the Head Start Early Learning Outcomes Framework

(continued)

Lesson	**P-LIT-2.** Child demonstrates an understanding of how print is used (functions of print) and the rules that govern how print works (conventions of print).	**P-LIT 3.** Child identifies letters of the alphabet and produces correct sounds associated with letters.	**P-LIT 6.** Child writes for a variety of purposes using increasingly sophisticated marks.	**P-MATH 3.** Child understands the relationship between numbers and quantities.	**P-MATH 4.** Child compares numbers.	**P-MATH 5.** Child associates a quantity with written numerals up to 5 and begins to write numbers.	**P-MATH 8.** Child measures objects by their various attributes using standard and non-standard measurement. Uses differences in attributes to make comparisons.	**P-MATH 9.** Child identifies, describes, compares, and composes shapes.	**P-SCI 1.** Child observes and describes observable phenomena (objects, materials, organisms, and events).	**P-SCI 2.** Child engages in scientific talk.	**P-SCI 3.** Child compares and categorizes observable phenomena.	**P-SCI 4.** Child asks a question, gathers information, and makes predictions.	**P-SCI 5.** Child plans and conducts investigations and experiments.	**P-SCI 6.** Child analyzes results, draws conclusions, and communicates results.
Nature Walks														
Animal Walk			•	•	•				•	•		•	•	•
Looking for Birds	•		•	•	•				•	•		•	•	•
Nature Bracelets	•	•		•					•	•	•	•		•
Finding Weeds	•	•	•		•		•		•	•	•	•		•
Branch Puzzles	•	•	•		•		•		•	•	•	•		•
Adopt a Tree		•	•	•	•		•		•	•	•	•		•
Comparing Leaves		•	•		•			•	•	•	•	•		•
Scavenger Hunt		•		•				•	•	•	•			•

Appendix B

Lesson Connections to the *Next Generation Science Standards*

Lesson	Asking Questions and Defining Problems	Developing and Using Models	Planning and Carrying Out Investigations	Analyzing and Interpreting Data	Using Mathematics and Computational Thinking	Constructing Explanations and Designing Solutions	Engaging in Argument From Evidence	Obtaining, Evaluating, and Communicating Information	**LS1.A**: Structure and Function	**LS1.B**: Growth and Development of Organisms	**LS1.C**: Organization for Matter and Energy Flow in Organisms	**LS1.D**: Information Processing	**LS2.A**: Interdependent Relationships in Ecosystems	**LS3.A**: Inheritance of Traits	**LS3.B**: Variation of Traits	**LS4.D**: Biodiversity and Humans	Patterns	Cause and Effect	Scale, Proportion, and Quantity	Systems and System Models	Energy and Matter: Flows, Cycles, and Conservation	Structure and Function	Stability and Change of Systems
Animals																							
Roly-Polies		•	•	•	•	•	•	•	•			•						•	•	•		•	
Jumping Crickets					•	•		•	•			•							•	•		•	
Observing Earthworms	•		•		•			•	•						•			•	•	•		•	
Snails			•	•	•			•	•			•			•			•	•	•		•	
Swimming Fish			•	•	•			•	•		•				•					•		•	
Critter Camouflage			•	•	•	•		•	•							•		•				•	
Spiderwebs			•		•			•	•		•						•					•	
Feeding Birds			•	•	•	•		•			•					•		•	•				
Plants																							
Sorting Seeds			•		•			•	•										•				
Seeds in Our Food	•		•	•	•	•	•	•	•				•				•	•	•		•	•	
Where Vegetables Come From			•		•	•	•	•	•														
Sprouting Seeds			•	•	•	•	•	•	•	•	•		•				•	•	•			•	
Terrariums			•	•	•			•			•		•			•			•				•
Smelling Plants			•		•			•								•			•				
Pumpkin Outsides	•		•	•	•			•	•						•				•				
Pumpkin Insides			•		•			•	•										•	•			

Lesson Connections to the *Next Generation Science Standards* (continued)

Lesson	Asking Questions and Defining Problems	Developing and Using Models	Planning and Carrying Out Investigations	Analyzing and Interpreting Data	Using Mathematics and Computational Thinking	Constructing Explanations and Designing Solutions	Engaging in Argument From Evidence	Obtaining, Evaluating, and Communicating Information	**LS1.A**: Structure and Function	**LS1.B**: Growth and Development of Organisms	**LS1.C**: Organization for Matter and Energy Flow in Organisms	**LS1.D**: Information Processing	**LS2.A**: Interdependent Relationships in Ecosystems	**LS3.A**: Inheritance of Traits	**LS3.B**: Variation of Traits	**LS4.D**: Biodiversity and Humans	Patterns	Cause and Effect	Scale, Proportion, and Quantity	Systems and System Models	Energy and Matter: Flows, Cycles, and Conservation	Structure and Function	Stability and Change of Systems
Nature Walks																							
Animal Walk			•	•	•	•		•			•					•	•	•					
Looking for Birds	•		•		•			•			•					•	•		•				
Nature Bracelets			•		•			•	•							•			•			•	
Finding Weeds	•		•	•				•	•							•	•					•	
Branch Puzzles	•	•	•	•	•	•	•	•	•								•		•	•			
Adopt a Tree			•		•	•	•	•	•								•		•				•
Comparing Leaves			•	•				•	•						•				•			•	
Scavenger Hunt			•					•	•							•						•	

Appendix C

The *A Head Start on Science* Program

Housed in the National Center for Science in Early Childhood at California State University, Long Beach, the *A Head Start on Science* (*HSOS*) program strives to provide quality science resources and professional development for early childhood educators. These efforts work to advance teachers' competence and comfort with pedagogical strategies that support children's understanding of science as well as to promote teachers' own appreciation and love of science. *HSOS* provides professional development institutes in California and conference presentations and workshops throughout North America. More information can be found at the program's website, *sci4kids.org*.

Appendix D

William C. Ritz, Founder of *A Head Start on Science*

Twenty years ago, few science education researchers were focused on early childhood. One of these few was Dr. William C. Ritz. In 1996, with support from the U.S. Department of Health and Human Services, he developed the *A Head Start on Science (HSOS)* curriculum and professional development for teachers of young children. Inspired by Rachel Carson's wonderful book (1956), Dr. Ritz decided on "encouraging a sense of wonder" as the theme for the *HSOS* project and set about encouraging teachers to engage children in activities that foster their curiosity and spark exploration and discovery through science. His tireless work resulted in an 89-activity curriculum guide for teachers that outlines the scientific practices embedded within the activity and includes instructions for facilitating those practices as well as follow-up activities to integrate learning in classrooms and with families. The *HSOS* curriculum guide was published by NSTA Press in 2007. This book has been widely disseminated, with over 18,000 copies sold, and was awarded the Distinguished Achievement Award by the Association of Educational Publishers. In 2001, with funding from the Honda Foundation, Dr. Ritz established 21 national *HSOS* training and dissemination sites throughout the United States. In 2008, he collaborated to help establish *HSOS* field center at the China National Institute for Educational Research in Beijing, and a Mandarin translation of the teachers' guide was developed. Worldwide, hundreds of *HSOS* workshops have been conducted, and thousands of preschool and primary-grade teachers have received the training, support, and encouragement that *HSOS* offers. The inspiring work of Dr. Ritz has brought meaningful science experiences to countless young learners. At 86 years young, Dr. Ritz continues to be an integral part of the *HSOS* project and remains an inspiration to us all.

Appendix E

Glossary of Key Science Terms

Animal. A multicellular organism that consumes other organisms in order to survive, has organs that can respond to stimuli, and can move voluntarily

Antennae. Paired appendages at the head of some animals used as sense organs, typically to determine touch, vibration, or odor

Bark. A protective layer lining the outside of woody stems and branches

Behavior. The way something acts; what something does

Bill. The hard, external part of a bird's mouth used to get food

Branch. A woody extension of a tree that grows out from the trunk or other branches, typically sprouting leaves at its tip and along its length

Camouflage. Coloring that allows animals to hide by making them look like their surroundings

Cerci. Paired appendages found at the rear of many insects (and other animals), typically used as a sensory organ, but in some species used to pinch or grab

Characteristic. A distinguishing trait or quality

Chlorophyll. A green pigment responsible for the absorption of light initiating photosynthesis

Clitellum. A smooth, raised band around an earthworm's body that contains the reproductive organs

Compare. To examine looking for similarities (or differences) between two or more objects/phenomena

Compound Eye. A type of eye consisting of many small, light-sensing visual units (i.e., many small eyes)

Data. Measurements or other information used for analyses or comparisons

Dissect. To carefully take apart a whole in order to examine the individual parts

Environment. The conditions that surround and affect an animal or other living thing (includes both living and nonliving/physical factors)

Experiment. To try out or test ideas; to discover, prove, or disprove something

Feather. A light, flat structure that grows from and covers the skin of a bird

Fin. A muscular organ attached to the body of fish that is used for propulsion, steering, or balance

Flower. The reproductive structure of many plants, typically consisting of reproductive organs (stamen and pistil) surrounded by brightly colored petals and green sepals

Foot. A large muscle on the underside of a snail's body that propels the snail; more generally, a body part that contacts the ground (or other surface) and helps with locomotion

Fruit. The seed-bearing structure of flowering plants, often adapted to aid in the dispersal of seeds—commonly known as edible plant parts that contain seeds

Germinate. The process by which the embryo within a seed sprouts and becomes a seedling

Graph. To make a diagram that shows the relationship between two or more variables

Investigate. Explore with an open mind to understand the world around you and objects, events, phenomena, experiences, etc.

Leaf. A flattened and typically green plant structure that is attached to a stem and serves as the primary site of photosynthesis

Legs. Paired appendages used by animals to walk, crawl, climb, etc.

Map. A diagram that represents and shows key features of an area

Measure. Describe the size, amount, or degree of something by using an instrument marked in standard units (e.g., a ruler) or by comparing it with an object of known size, amount, or degree

Mucus. A bodily secretion that lubricates a surface; in snails, it becomes a thin layer covering the foot and significantly reducing friction, thereby helping the snail to move (relatively) quickly and safely

Observe. Using our senses to gather information about specific objects/phenomena in the world around us and recording that information

Operculum. A hard plate, readily visible at the rear of the head of many fish, including goldfish, that covers and protects the gills

Ovipositor. An organ extending from the abdomen of female insects typically used to deposit fertilized eggs

Petal. Modified leaves within flowers that are typically brightly colored and serve to attract pollinators

Photosynthesis. A process used by plants and other organisms to convert light energy into chemical energy

Predator. An animal that captures and eats another animal for food

Prey. An animal that is hunted and eaten by another animal for food

Pumpkin. A large, typically round and orange, many-seeded fruit related to squash

Root. A (typically underground) part of a plant that functions in the absorption of water and nutrients

Scent. A specific odor or fragrance that is emitted by something; a smell

Seasons. One of the four periods (spring, summer, autumn, and winter) into which the year is commonly divided, each with characteristic weather and daylight hours

Seed. A small "baby" plant enclosed within a seed coat

Seed Coat. The outer part of the seed

Seedling. A small, recently germinated plant

(**Body**) **Segment**. The individual, and often repetitive, sections of the body of some organisms

Set. A collection of objects considered as a whole

Setae. Stiff hairs that extend from an earthworm's body and aid in movement

Shell. The external skeleton of snails and many other animals, typically used to protect from predation, physical damage, and dehydration

Sorting. The act of grouping similar objects, ideas, or phenomena

Spider. An eight-legged invertebrate (animal without a backbone) related to ticks and scorpions

Stem. The primary plant body, typically growing upright and aboveground and supporting leaves and other structures

Tail. An appendage at the rear of an animal that typically aids in propulsion or balance; in birds, tails are feathered and serve to help maintain balance, especially in flight

Tail fin. Fin found at the rear of a fish's body and used to help propel the fish forward; also called the caudal fin

Tally. Making a mark for each time something happens

Terrarium. A clear container for growing and observing plants indoors

Tree. A long-lived, woody plant, typically having a single stem or trunk, growing to a considerable height, and bearing lateral branches at some distance from the ground

Trunk. The main woody stem of a tree that typically grows upright and accounts for much of a tree's height

Twig. A small branch often lacking leaves

Vegetable. Edible plant parts that can be derived from various parts of a plant, including roots, stems, leaves, flowers, and fruits

Web. A network of silk threads woven by a spider

Weed. A typically fast-growing plant that is able to disperse seeds to and grow in many different places, including places it may not be wanted

Wing. Paired appendages that enable some insects (and other animals) to fly

References

Aber, L. 2002. *Grandma's button box.* Minneapolis: Kane Press.

Adams, J. 2011. *Fruits.* New York: Rosen Publishing.

Alderfer, J. 2013. *Bird guide of North America.* Washington, DC: National Geographic.

Aliki. 2015. *My five senses.* New York: HarperCollins.

Allen, J., and T. Humphries. 2003. *Are you a spider?* London: Kingfisher.

Allen, J., and T. Humphries. 2013. *Are you a snail?* London: Kingfisher.

Anthony, J. 1997. *The dandelion seed.* Nevada City, CA: Dawn Publications.

Arnold, C. 2009. *Wiggle and waggle.* Watertown, MA: Charlesbridge.

Arnold, K. 2002. *Let's find it: My first nature guide.* New York: HarperCollins.

Arnosky, J. 1992. *Crinkleroot's guide to knowing the birds.* New York: Simon and Schuster.

Arnosky, J. 1993. *Crinkleroot's 25 birds every child should know.* New York: Simon and Schuster.

Arnosky, J. 2000. *I see animals hiding.* New York: Scholastic.

Atkin, J. M., and R. Karplus. 1962. Discovery or invention? *The Science Teacher* 29 (5): 45–51.

Axtell, D. 2015. *We're going on a lion hunt.* New York: Two Lions.

Ayers, K. 2008. *Up, down, and around.* Cambridge, MA: Candlewick.

Ayers, P., and G. Percy. 1988. *When Dad cuts down the chestnut tree.* New York: Knopf.

Bash, B. 1992. *Urban roosts: Where birds nest in the city.* New York: Little, Brown.

Beals, S. 2011. *Nests.* San Francisco: Chronicle.

Bennett, K. 2008. *Not Norman.* Cambridge, MA: Candwick.

Berger, C. 2008. *The little yellow leaf.* New York: Greenwillow.

Berger, M. 1998. *Chirping crickets.* New York: HarperCollins.

Berger, M. 2003. *Spinning spiders.* New York: HarperCollins.

Bernard, R. 2001. *Tree for all seasons.* Washington, DC: National Geographic.

Bishop, C. 2016. *Why do plants have fruits?* New York: Rosen Publishing.

Bishop, N. 2012. *Spiders.* New York: Scholastic.

Bloom, V. 1997. *Fruits: A Caribbean counting poem.* New York: MacMillion.

Bodach, V.K. 2007. Fruits. New York: First Avenue Editions.

Bodach, V.K. 2016. Seeds. New York: First Avenue Editions.

Brendler, C. 2009. *Winnie Finn, worm farmer.* Dongguan City, Guangdong Province: South China Printing.

Brisson, P. 2000. *Wanda's roses.* Honesdale, PA: Boyd's Mill Press.

Brocklehurst, R. 2010. *1001 animals to spot.* London: Usbourne.

Brown, P. 2009. *The curious garden.* New York: Hachette.

Browne, E. 1999. *Handa's surprise*. Cambridge, MA: Candwick.

Bruning, M. 2007. *Sorting by color.* Mankato, MN: Capstone.

Bulla, C. R. 2016. *A tree is a plant.* New York: HarperCollins.

Bunting, E. 2000. *Flower garden.* New York: HMH Books for Young Readers.

Burts, D., K. Berke, C. Heroman, H. Baker, T. Bickart, P. Tabors, and S. Sanders. 2016. *Teaching strategies GOLD: Objectives for development and learning: Birth through third grade.* Bethesda, MD: Teaching Strategies, LLC.

Cabell, S. Q., J. DeCoster, J. LoCasale-Crouch, B. K. Hamre, and R. C. Pianta. 2013. Variation in the effectiveness of instructional interactions across preschool classroom settings and learning activities. *Early Childhood Research Quarterly* 28 (4): 820–830.

Carle, E. 1997. *The very quiet cricket.* New York: Penguin.

Carle, E. 2007. *My very first book of animal homes.* New York: Penguin.

Carretero, M. 2011. *Roly-polies.* Madrid, Spain: Graphicas AGA.

Carson, R. 1956. *The sense of wonder*. New York: Harper and Row Publishers.

Castella, K. and B. Boyl. 2006. *Discovering nature's alphabet.* Berkeley, CA: Heyday Books.

Caudill, R. 2004. *A pocketful of cricket.* New York: Henry Holt.

Chernesky, F. S. 2013. *Pick a circle, gather squares.* New York: Albert Whitman.

Child, L. 2003. *I will never not ever eat a tomato.* Cambridge, MA: Candlewick.

Cole, H. 1998. *I took a walk*. New York: Greenwillow.

Cole, H. 2003. *On the way to the beach.* New York: Greenwillow.

Collard, S. 2002. *Beaks!* Watertown, MA: Charlesbridge.

Coughlan, C. 1998. *Crickets.* Mankato, MN: Capstone Press.

Cronin, D. 2003. *Diary of a worm.* New York: HarperCollins.

Dickmann, N. 2012. *Vegetables.* New York: Heinemann.

Diesen, D. 2013. *The pout-pout fish*. New York: Farrar.

DK Publishing. 1999. *Scratch and sniff: Food*. New York: DK Preschool.

DK Publishing. 1999. *Scratch and sniff: Garden.* New York: DK Preschool.

Duke, K. 2012. *Ready for pumpkins.* New York: Knopf.

Ehlert, L. 1996. *Eating the alphabet*. Orlando: Harcourt.

Ehlert, L. 2003. *Planting a rainbow*. New York: HMH Books for Young Readers.

Emmett, J. 2009. *Leaf trouble*. New York: Scholastic.

Esbaum, J. 2009. *Seed, sprout, pumpkin, pie*. Washington, DC: National Geographic.

Fan, T., and E. Fan. 2016. *The night gardener*. New York: Simon and Schuster.

Farmer, J. 2004. *Pumpkins*. Watertown, MA: Charlesbridge.

Fleming, D. 1995. *In the tall, tall grass*. New York: Square Fish.

Fowler, A. 1999. *A snail's pace*. New York: Children's Press.

Franchino, V. 2015. *Animal camouflage*. New York: Scholastic.

Franco, B. 2007. *Birdsongs*. New York: Margaret K. McElderry Books.

French, V. 1995. *Oliver's vegetables*. London: Hachette.

French, V. 1998. *Oliver's fruit salad*. London: Hachette.

French, V. 2012. *Yucky worms*. Somerville, MA: Candlewick.

Galbraith, K. 2015. Planting the wild garden. Atlanta: Peachtree.

Garelick, M. 1995. *What makes a bird a bird?* New York: Mondo.

Garland, S. 2009. *Eddie's garden and how to make things grow*. London: Frances Lincoln.

Gelman, R., K. Brenneman, G. Macdonald, and M. Roman. 2009. *Preschool pathways to science: Facilitating scientific ways of thinking, talking, doing, and understanding*. Baltimore: Brookes Publishing.

Gianferrari, M. 2016. *Coyote moon*. New York: Roaring Brook Press.

Gibbons, G. 1988. *The seasons of Arnold's apple tree*. New York: HMH Books for Young Readers.

Gibbons, G. 1991. *From seed to plant*. New York: Holiday House.

Gibbons, G. 1993. *Spiders*. New York: Holiday House.

Gibbons, G. 1999. *The pumpkin book*. New York: Holiday House.

Gibbons, G. 2002. *Tell me, tree: All about trees for kids*. New York: Little, Brown.

Gibbons, G. 2008. *The vegetables we eat*. New York: Holiday House.

Gibbons, G. 2016. *The fruits we eat*. New York: Holiday House.

Glaser, L. 1994. *Wonderful worms*. Minneapolis: Millbrook.

Glaser, M. 2009. *Singing crickets*. Minneapolis: Milbrook.

Gomez Zwiep, S., and W. J. Straits. 2013. Inquiry science: The gateway to English language proficiency. *Journal of Science Teacher Education* 24 (8): 1315–1331.

Grindley, S. 2013. *Snail trail*. New York: Hachette.

Hall, Z. 1999. *It's pumpkin time*. New York: Scholastic.

Hall, Z. 2000. *Fall leaves fall*. New York: Scholastic.

Hansen, G. 2016. *Fruits*. Minneapolis: Abdo.

Helbrough, E. 2009. *1001 bugs to spot*. London: Usbourne.

Heller, R. 1992. *How to hide a butterfly and other insects*. New York: Grosset and Dunlap.

Henkes, K. 2010. *My garden*. New York: Greenwillow.

Hickman, P. 1997. *A seed grows*. Toronto: Kids Can Press.

Hicks, B. J. 2014. *Monsters don't eat broccoli*. New York: Dragonfly.

HighScope Educational Research Foundation. 2015. *Preschool child observation record–COR advantage*. Ypsilanti, MI: HighScope Press.

Himmelman, J. 1999. *A dandelion's life*. New York: Children's Press.

Himmelman, J. 2000. *A pill bug's life*. New York: Children's Press.

Himmelman, J. 2001. *An earthworm's life*. New York: Children's Press.

Hoberman, M. A. 2000. *The seven silly eaters*. Orlando: Harcourt.

Holub, J. Pumpkin countdown. New York: Albert Whitman.

Hopkins, H. J. 2013. *The tree lady*. New York: Simon and Schuster.

Howard, K. 2004. *Little bunny follows his nose*. New York: Random House.

Hubbell, W. 2000. *Pumpkin jack*. New York: Albert Whitman.

Jeffers, O. 2009. *The great paper caper*. New York: Penguin.

Jeffers, O. 2011. *Stuck*. New York: Penguin.

Jenkins, S. 2012. *The beetle book*. New York: Harcourt.

Jenson-Elliott, C. 2014. *Weeds find a way*. New York: Beach Lane.

Johnson, S. 1999. *Alphabet city*. New York: Puffin Books.

Jonas, A. 1999. *Bird talk*. New York: Greenwillow.

Jordan, H. J. 2015. *How a seed grows*. New York: Harper Collins.

Kalman, B. 2007. *What shape is it?* New York: Crabtree.

Karwoski, G. L. 2013. *Thank you, trees!* Minneapolis: Lerner Classroom.

Kelly, I. 2009. *Even an ostrich needs a nest*. New York: Holiday House.

Kite, P. 1998. *Dandelion adventures*. Minneapolis: Millbrook.

Konicek-Moran, R., and K. Konicek-Moran, 2017. *From flower to fruit*. Arlington VA: NSTA Press.

Kudlinski, K. 1999. *Dandelions*. Minneapolis: Lerner Classroom.

Lauber, P. 1994. *Be a friend to trees*. New York: HarperCollins.

Lawrence, E. 2012. From bird poop to wind: How seeds get around. New York: Bearport Publishing.

Lee, J. 2016. *See it grow: Pumpkin*. New York: Bearport Publishing.

Levenson, G. 2002. *Pumpkin circle: The story of a garden*. San Fransico: Tricycle.

Lewin, T. 2015. *Can you see me?* New York: Holiday House.

Lewis, K. 2005. *The runaway pumpkin*. New York: Scholastic.

Llewellyn, C. 1997. *Spiders have fangs*. New York: Copper Beech.

Locker, T. 2001. *Sky tree*. New York: HarperCollins.

Loewen, N. 2005. *Garden wigglers*. Minneapolis: Picture Window Books.

Lowery, L. F. 2015. *Animals two by two.* Arlington, VA: NSTA Press.

Lundblad, K., and B. Kalman. 2005. *Animals called fish.* New York: Crabtree.

Lyons, D. 2002. *The tree.* Kirkland, WA: Illumination Arts.

Macken, J. E. 2008. *Flip, float, fly.* New York: Holiday House

Maestro, B. 2000. *How do apples grow?* New York: HarperCollins.

Maestro, B. 2015. *Why do leaves change color?* New York: HarperCollins.

Mariconda, B. 2008. *Sort it out.* Mount Pleasant, SC: Arbordale Publishing.

Marks, J. 2012. *Sorting by size.* Mankato, MN: Capstone.

Martin, D. 1999. *Animal homes.* London: Usborne.

Mashburn, A. J., R. C. Pianta, B. K. Hamre, J. T. Downer, O. A. Barbarin, D. Bryant, M. Burchinal, D. M. Early, and C. Howes. 2008. Measures of classroom quality in prekindergarten and children's development of academic, language, and social skills. *Child Development* 79 (3): 732–749.

Matheson, C. 2013. *Tap the magic tree.* New York: Greenwillow.

Mazzola, F. 1997. *Counting is for the birds.* Watertown, MA: Charlesbridge.

McDonnell, M. R. 2013. *Sorting snakes.* New York: Gareth Stevens.

McKelvey, K. 2015. *Dandelions.* Wollombi, Australia: Exisle Publishing.

McNamara, M. 2007. *How many seeds in a pumpkin?* New York: Schwartz & Wade.

McQuinn, A. 2017. *Lola plants a garden.* Watertown, MA: Charlesbridge.

Metzger, S. 2008. *We're going on a leaf hunt.* New York: Scholastic.

Milbourne, A. 2014. *Peek inside animal homes.* London: Usborne.

Mora, P. 1994. *Pablo's tree.* New York: Simon and Schuster.

Morgan, E. 2013. *Next time you see a pill bug.* Arlington, VA: NSTA Press.

Morgan, E. 2014. *Next time you see a maple seed.* Arlington, VA: NSTA Press.

Morgan, E. 2015. *Next time you see a spiderweb.* Arlington, VA: NSTA Press.

Moulton, M. K. 2010. *The very best pumpkin.* New York: Simon and Schuster.

Muldrow, D. 2016. *We planted a tree.* New York: Dragonfly.

Murphy, M. 2013. *Slow snail.* Cambridge, MA: Candlewick.

Murray, D. 2016. *City shapes.* New York: Little, Brown.

Murray, L. 2016. *Spiders.* Mankato, MN: Creative Paperbacks.

National Association for the Education of Young Children. 2015. *NAEYC early childhood program standards and accreditation criteria. www.naeyc.org/academy.*

National Governors Association Center for Best Practices and Council of Chief State School Officers (NGAC and CCSSO). 2010. *Common core state standards.* Washington, DC: NGAC and CCSSO.

Nettleton, P. H. 2004. *Look, listen, taste, touch, and smell: Learning about your five senses* Mankato, MN: Picture Window Books

NGSS Lead States. 2013. *Next Generation Science Standards: For states, by states*. Washington, DC: National Academies Press. www.nextgenscience.org/next-generation-science-standards.

Numeroff, L. 2012. *It's pumpkin day, mouse!* New York: Balzer and Bray.

Olson, G. 2011. *My plate and you*. Mankato, MN: Capstone.

Oppenheim, J. 1988. *Have you seen birds?* New York: Scholastic.

Otto, C. 1996. *What color is camouflage?* New York: Harper Collins.

Pallotta, J. 1989. *The flower alphabet book*. Watertown, MA: Charlesbridge.

Pallotta, J. 2010. *Who will plant a tree?* Mankato, MN: Sleeping Bear Press.

Pascoe, E. 2001. *Birds use their beaks*. New York: Gareth Stevens.

Pfeffer, W. 2003. *Wiggling worms at work*. New York: Harper Collins.

Pfeffer, W. 2015. *From seed to pumpkin*. New York: HarperCollins.

Pfeffer, W. 2015. *What's it like to be a fish?* New York: HarperCollins.

Pfister, M. 1999. *The rainbow fish*. Zurich, Switzerland: North-South Books.

Posada, M. 2000. *Dandelions: Stars in the grass*. New York: Scholastic.

Previn, S. 2014. *Find Spot!* New York: Little, Brown.

Rau, D. M. 2006. *Star in my orange: Looking for nature's shapes*. New York: First Avenue Editions.

Richards, J. 2006. *A fruit is a suitcase for seeds*. New York: First Avenue.

Ritz, W., ed. 2007. *A head start on science: Encouraging a sense of wonder*. Arlington, VA: NSTA Press.

Robbins, K. 1998. *Autumn leaves*. New York: Scholastic.

Robbins, K. 2005. *Seeds*. New York: Atheneum Books

Rockwell, A. 1999. *One bean*. New York: Bloomsbury.

Rockwell, L. 2009. *Good enough to eat*. New York: Holiday House.

Rodman, M. A. 2009. *A tree for Emmy*. Atlanta: Peachtree.

Roemer, H. B. 2006. *What kind of seeds are these?* New York: Cooper Square Publishing.

Rosen, M. 1997. *We're going on a bear hunt*. New York: Simon and Schuster.

Rustad, M. 2011. *Fall leaves: Colorful and crunchy*. New York: Millbrook Press.

Rustad, M. 2011. *Fall pumpkins, orange and plump*. New York: Millbrook Press.

Ryan, P. M. 2001. *Hello ocean*. Watertown, MA: Charlesbridge.

Ryder, J. 1988. *The snail's spell*. New York: Puffin Books.

Savage, S. 2013. *Ten orange pumpkins*. New York: Dial Books

Sayre, A. P. 2014. *Rah, rah, radishes*. New York: Simon and Schuster.

Sayre, A. P. 2016. *Squirrels leap, squirrels sleep*. New York: Henry Holt.

Schuette, S. 2003. *An alphabet salad*. Mankato, MN: Capstone Press.

Schuh, M. 2011. *Apples grow on a tree*. Mankato, MN: Capstone Press.

Schuh, M. 2012. *Fruits on my plate*. Mankato, MN: Capstone Press.

Schuh, M. 2012. *Vegetables on my plate*. New York: Lerner.

Schwartz, D. 2013. *Rotten pumpkin: A rotten tale in 15 voices*. New York: Creston Books

Schwartz, D., and Y. Schy. 2011. *Where in the wild?* New York: Berkeley, CA: Tricycle.

Sheehan, K. 2014. *The dandelion's tale*. New York: Schwartz & Wade.

Sidman, J. 2011. *Swirl by swirl*. New York: HMH Books.

Simon, C. 2011. *Pumpkin fever*. New York: Children's Press

Sill, C. 2013. *About birds: A guide for children*. Atlanta: Peachtree.

Sill, C. 2017. *About fish: A guide for students*. Atlanta: Peachtree.

Snow, V. B. 2013. *Fall walk*. New York: Gibbs Smith.

Spurr, E. 2012. *In the garden*. Atlanta: Peachtree.

Squire, A. 2002. *Animal homes*. New York: Children's Press.

Stein, D. E. 2010. *Leaves*. New York: GP Putnam's and Sons.

Stewart, M. 2014. *Feathers: Not just for flying*. Watertown, MA: Charlesbridge.

Stewart, S. 2007. *The gardener*. New York: Square Fish.

Stockdale, S. 2012. *Fabulous fishes*. Atlanta: Peachtree.

Stockade, S. 2013. *Bring on the birds*. Atlanta: Peachtree.

Stokes, D. 1996. *Stokes beginner's guide to birds*. New York: Little Brown.

Swinburne, S. 1999. *Unbeatable beaks*. New York: Henry Holt.

Taback, S. 2009. *City animals*. New York: Blue Apple Books.

Tagliaferro, L. 2004. *Spiders and their webs*. Mankato, MN: Capstone.

Teckentrup, B. 2016. *Find the triangle*. New York: Sterling Children's Books.

Teckentrup, B. 2016. *Tree: A peek-through picture book*. New York: Doubleday.

Thornhill, J. 1996. *Wild in the city*. New York: Little, Brown.

Titherington, J. 1990. *Pumpkin pumpkin*. New York: Greenwillow Books.

Tokuda, Y. 2006. *I'm a pill bug*. La Jolla, CA: Kane Miller.

Ungerer, T. 2015. *Snail, where are you?* London: Phaidon Press.Urdy, J. 1987. *A tree is nice*. New York: HarperCollins.

U.S. Department of Health and Human Services. 2015. *Head start early learning outcomes framework: Ages birth to five*. Washington, DC: Administration for Children and Families.

Waldron, M. 2014. *Seeds and fruit*. New York: Heinemann.

Wallace, N. E. 2013. *Seeds! Seeds! Seeds!* New York: Two Lions.

Ward, J. 2014. *Mama built a little nest*. New York: Beach Lane.

Waxman, L. H. 2009. *Let's look at snails*. Minneapolis: Lerner Classroom.

Wellington, M. 2007. *Zinnia's flower garden*. New York: Puffin Books.

Wellington, M. 2015. *My leaf book*. New York: Dial Books.

Willis, J. 2009. *The bog baby*. New York: Schwartz & Wade.

Winter, J. 2007. *The tale of pale male*. Orlando: Harcourt.

Winter, J. 2008. *Wangari's trees of peace: A true story from Africa*. Orlando: Harcourt.

Wood, A. 1997. *Birdsong*. Orlando: Harcourt.

Yolen, J., and H. Stemple. 2015. *You nest here with me*. Honesdale, PA: Highlights.

Index